B. P. Pratten

Half hours with the animals

Narratives exhibiting thought, sympathy and affection in the brute creation

B. P. Pratten

Half hours with the animals
Narratives exhibiting thought, sympathy and affection in the brute creation

ISBN/EAN: 9783337228682

Printed in Europe, USA, Canada, Australia, Japan

Cover: Foto ©berggeist007 / pixelio.de

More available books at **www.hansebooks.com**

HALF HOURS
WITH THE ANIMALS

NARRATIVES

EXHIBITING

THOUGHT, SYMPATHY, AND AFFECTION

ILLUSTRATIONS.

THE following pages claim to be nothing more than a collection of facts and incidents bearing upon a particular feature and condition of the Animal World, and intended mainly to enlarge the reader's knowledge. The book does not aim to maintain any principle or moral system by argument. Nothing, it is believed, will be found in it but simple truth—that is, nothing but narratives given by eye-witnesses, or otherwise supported by adequate testimony. And when thus placed before the reader's mind, those facts are left to produce their natural effect, without being applied to establish any theory, or promote any opinion.

Yet it can hardly be that they will fail, in one respect, to do good. For their natural tendency is, to raise some of God's creatures in our estimation, and to excite feelings of sympathy towards them; and so to lessen the distance which has hitherto existed between man and the animals which surround him. Hitherto these animals have been too much regarded as mere slaves— as little better than machines—and have thus excited no livelier feelings than those with which we regard the steam-engine or the plough. But when, by study and thought, we come to perceive that some of these creatures can reason, can sympathise, and can form strong attachments, their position with reference to mankind must change, must be elevated; and thus, in the simplest and easiest way, "Cruelty to Animals" must grow more odious and more irrational.

So far, without controversy, good may be done. Perhaps, in another respect, a thoughtful consideration of the facts here exhibited may have a beneficial influence. We shall merely indicate it, and then leave the subject with our readers.

The fact, surely, is worthy of some thought and investigation, that there are few even of the wildest and most savage of the creatures that cannot be brought into friendship with man, if he chooses to endeavour to gain their good will. Even the elephant, the wisest and mightiest of the brute creation,—he who wants so little at the hands of man,—may easily be made a sympathizing companion. Nor must we forget those still ruder creatures, the lion and the bear, both of whom have been brought into amicable relations. In fact, we are hardly justified in supposing that there is any one of the animals of so irreclaimable a disposition as to reject absolutely every overture that man could make, to establish a good understanding.

This thought naturally leads to another. We are told in God's Word that a time is coming when "the wolf shall dwell with the lamb, and the leopard shall lie down with the kid, and the calf and the young lion and the fatling together, and a little child shall lead them." Doubtless this language may be figurative, and it may be figurative *only*. But equally allowable is it to believe, that it may be figurative, and *literal also*. It will hardly be doubted that, whenever the time shall come in which the hitherto ferocious and savage among mankind shall be brought into Christian habits of thought and feeling, the animal world will be immensely benefited by the change. The horse will no longer be the ill-used slave he too often is at present; the dog will no longer be taught and trained to chase, and worry, and tear in pieces creatures whose right to live is equal to his own; the dove and the hare will no longer be made the mere victims they now are;—in short, cruelty will be banished from among men, and the only question will be, how these beautiful creatures of God's hand can be raised and rendered happier and better than they have yet been. The thought, if pursued, would lead us far; but we must restrain ourselves, remembering the principle with which we started;—that our object was, not to advance theories, or promote systems, but simply to bring together some useful and not uninteresting information.

Yet we cannot forget the apostle's words, which are clearly pregnant with deep and important meanings, "The *whole creation* groaneth and travaileth in pain together until now." "The earnest expectation of the creature waiteth for the manifestation of the sons of God."

As we have already said, we have no thought of propounding a theory—a system—but we cannot doubt that Paul's words are true, are certain, and shall be fulfilled. There will be, in the appointed time, a "manifestation," a coming forth into the light, of the sons of God, of those who have obtained entrance into

His family. Of the nature, the time, and the character of that entrance, we cannot now say anything. It would require more space than we can give. But it will be seen—will be so seen as to be no matter of doubt or question. And when that day shall come, the "earnest expectation of the creature" shall be fully met: "The wolf and the lamb shall feed together, and the lion shall eat straw like the bullock, and dust shall be the serpent's meat." These are true words, and there are various signs which seem to tell us that the dawn at least of that day is appearing.

HALF HOURS WITH THE ANIMALS.

HALF HOURS WITH THE ANIMALS.

THE subject dealt with in the following pages is one of great interest, and wonderful variety. A single topic divides itself, in real life, into a thousand parts.

No one doubts the vast superiority of the human race over "the brute creation"—"the beasts that perish." First of all, in the *extent* of the reasoning power; but more obviously in the possession of speech and written language. In these respects man is placed, perhaps, as far above the animals as the angels are above man.

But, after fully stating this at the outset, we must next advert to a few distinct points, in which some members of the brute creation show themselves in possession of certain faculties and powers, and, more, of feelings and affections, far exceeding those which are commonly exhibited by human beings.

It is perhaps in the language of hyperbole that Sydney Smith says,—

"It would take a senior wrangler at Cambridge ten hours a day for three years together, to know enough mathematics for the calculation of those problems with which not only every queen bee, but every undergraduate grub, is acquainted the moment it is born."

This, we repeat, may seem hyperbolical ; but the instances are numerous, and most certain, in which dogs, carried, perhaps in enclosed carriages, for hundreds of miles, have, on being released, at once gone back direct, and in the shortest possible time, to the place from which they had been taken, though the roads must have been quite unknown to them. This, it is quite evident, is a feat which no human being could possibly equal.

The chief advantage, however, which is clearly possessed by some animals,—the dog being evidently the first in this respect,—consists in warmth and steadfastness of affection. Instances are numerous, and are constantly recurring, in which dogs have defended their masters at the hazard, or even by the sacrifice, of their own lives. Other cases, too, are well known, in which dogs have refused to live after their masters had departed; and not a few, in which they have taken up their abodes on the graves of those whom they loved, and have resolutely refused to quit. In many respects, it must be admitted, the depth of attachment shown by these creatures for those who have been kind to them, is far beyond anything that is known, except in very rare instances, among mankind.

But we must not anticipate. The following pages will consist entirely of narratives, and those narratives cannot but be interesting. No argument will be founded on them; no conclusions will be drawn. We desire only to collect and to put into an accessible form a number of remarkable facts, leaving the reader to deduce from them such lessons as he may. We will only add one caution, which seems called for by certain doubtful speculations already offered to the public: that it will be wise to remember, that nothing that we can do or say in this matter can amount to more than a surmise,— a supposition. We may collect and arrange facts; but of the conclusions to which they tend we can know little or nothing.

I.

IN THE HOUSE.

IT is natural that we should begin with that animal which stands nearest to man, and is most familiar with him. But this chapter will necessarily be limited in its extent, by a remembrance of the volume published a few months since,* the contents of which it would be obviously improper to reproduce. Still, many anecdotes remain, which are not found in that volume; and some of these we shall here insert, without any attempt at decoration, and without comment.

Washington Irving, in describing a visit to Abbotsford, says,—

"After my return from Melrose Abbey, Scott proposed a ramble, to show me something of the surrounding country. As we sallied forth, every dog in the establishment turned out to attend us.

"There was the old and well-known staghound, 'Maida,'—a noble animal, and a great favourite of

---

* "Dog-Life: Narratives exhibiting Instinct, Intelligence, Fidelity, Attachments, etc." London, 1875.

THE SENTINEL.

Scott's; and 'Hamlet,' the black greyhound,—a mild, thoughtless youngster, which had not yet arrived at years of discretion; and 'Finella,' a beautiful setter, with soft silken hair, long pendent ears, and a mild eye, —the parlour favourite. When in front of the house, we were joined by a superannuated greyhound, which came from the kitchen, wagging his tail, and was cheered by Scott as an old friend and comrade. In our walks, Scott would frequently pause in conversation, to notice his dogs and speak to them, as if rational companions; and, indeed, there appears to be a vast deal of rationality in these faithful attendants on man, derived from their close intimacy with him. 'Maida' deported himself with a gravity becoming his age and size, and seemed to consider himself called upon to preserve a great degree of dignity and decorum in our society. As he jogged along a little distance ahead of us, the young dogs would frolic about him, leap on his neck, worry at his ears, and endeavour to tease him into a gambol. The old dog would keep on for a long time, with imperturbable solemnity, now and then seeming to rebuke the wantonness of his young companions. At length, he would make a sudden turn, seize one of them, and tumble him in the dust; then, giving a glance at us, as much as to say, 'You see, gentlemen, I can't help giving way to this nonsense,' would resume his gravity, and jog on as before. Scott amused himself with these peculiarities. 'I make no doubt,' said he, 'when "Maida" is alone with these young dogs, he throws gravity aside, and plays the boy as much as any of them ; but he is ashamed to do so in our company, and seems to say, "Ha' done with your nonsense, youngsters : what will the laird and that other gentleman think of me if I give way to such foolery ? " '

"While we were discussing the humours and the peculiarities of our canine companions, some object provoked their spleen, and produced a sharp and petulant barking from the smaller fry; but it was some time before ' Maida' was sufficiently roused to ramp forward two or three bounds, and join the chorus with a deep-mouthed ' bow-wow.' It was but a transient outbreak, and he returned instantly, wagging his tail, and looking up dubiously in his master's face, uncertain whether he would receive censure or applause. 'Ay, ay, old boy,' cried Scott, ' you have done wonders ; you have shaken the Eildon Hills with your roaring; you may now lay by your artillery for the rest of the day. "Maida," ' continued he, ' is like the great gun of Constantinople : it takes so long to get it ready that the smaller guns can fire off a dozen times first ; but when it does go off, it does great mischief.' "

Further on we have a peep at Sir Walter's dinner-table. Irving says,—

"Around the table were two or three dogs in attendance. 'Maida,' the old staghound, took his seat at Scott's elbow, looking up wistfully in his master's eye; while ' Finella,' the pet spaniel, placed herself near Mrs. Scott,—by whom, I soon perceived, she was completely spoiled. The conversation happening to turn on the merits of his dogs, Scott spoke with great feeling and affection of his favourite terrier, ' Camp,' who is depicted by his side in the earlier engravings of him. He talked of him as of a real friend whom he had lost; and Sophia Scott, looking up archly in his face, observed that ' Papa shed a few tears when " Camp" died.' "

Mr. Grantley Berkeley writes,—

" When I settled at Beacon Lodge, I was perplexed, after a while, by some dog who hunted at night, and who killed not only many of my tame rabbits, but also some of my favourite bantams, who were sitting on their nests.

" One night the keeper came running up, and crying, ' The dog, sir! the dog!—there are two!' We ran to the cover where they were. In a moment the keeper fired, and struck one, without materially injuring him; but the other we could not see. He could only escape by one of two ways: one of these was by going right up to my window, and this I thought not likely; and the other was by the open field. I watched for him there; but no dog appeared. We both wondered what could

have become of a large black retriever, who had been
seen but a few minutes before.

"During this time my wife was in the dining-room,
the window of which opens upon the lawn. Suddenly
she was aware of a large black dog, sitting in an im-
ploring attitude, with his nose against the window-pane,
and who wagged his tail when he saw he had gained her
attention. She went to the window, and when she saw
what the dog wanted, she threw up the sash and admitted him; and thus, when I returned, I found the dog
we had been looking for lying quietly on the rug in the
drawing-room. Of course he received the hospitality he
had so sensibly and trustingly sought, and in a few days
was sent home to his master."

In *Science Gossip* we read,—

"There is a narrow Westminster street, with little
shops and lodging-houses on either side of it. It is a
dirty, noisy little street, and gives access to a broader,
quieter one, with better houses, whose faces look out into
the park and its green elm-trees. Troops of children
from the little street play and shout in, and ring the
door-bells of, the great one. From the windows of
one of those houses I have watched the games of the
children, and observed them to be generally shared
by an ugly, smooth, white dog, with a sharp nose and
a few black spots. When an organ-man came to play,

the children danced with the dog, holding his unresist·
ing fore-paws. If a woman from the little street came
through with her basket, on her way to market, off
started the dog, with barks and leaps of joy, to accompany her as her guard and companion, and returned
with her when business was over. We never could
make out to whom the dog belonged. We met him
sometimes with one person, sometimes with another.
All the children loved him, and the grown people seemed
to have a friend and possession in him. His name we
found to be 'Spot'; and one day we found out poor
'Spot's' private history. In the little street was a
very small sweet-shop, much favoured by the children
of our family, amongst whom it went by the name of
'The Little Woman's.' The little woman sold haber-
dashery and illustrated papers, besides her sweets; and
during his leisure hours 'Spot' was often to be found
sitting bolt upright on her doorstep. We used to stroke
his head as we passed him, but he would scarcely care
to recognise us. His mind was fully occupied with his
own friends; and kind friends they seem to have been,
too. First of all, however, came a tragedy. Some
cruel person half hung the poor dog, and cut his throat.
A kind woman and her daughter, living in the street,
took the dog in, sewed up his throat, nursed him carefully, and restored him to health. This seems to have
been the commencement of his career as the street dog ;

but, instead of his being homeless, the street itself owned him, and became his home. He slept at the little sweet-shop woman's ; and every day she bought a piece of meat of the cats'-meat man, so that ' Spot ' was sure of one meal. I have offered him a bit of biscuit some-times, when I met him, but he did not seem to care about eating it ; so I think he was well fed. The two streets harboured no other dogs ; for ' Spot ' would come tearing down the whole length of them, and clear out any strange dog who ventured to loiter there. For years he has been a loved and valued street dog. Every **one seemed to** speak kindly to him ; and I have met him long distances from home, following various masters and mistresses. He always looked business-like and decided. At length came the new rule about the dog-licenses. Of course, no one had ever paid a tax for ' Spot '—no one need claim to be his real master ; but the ' little woman ' thought differently about the license. As a street dog—an ordinary, vulgar street dog—poor ' Spot ' might have become the prey of the police ; so this good woman went the round of the other little houses and shops, and collected a little here and there from ' Spot's ' kind friends, until she had enough to pay the license. So the street keeps its own dog with its own license. I have left the neigh-bourhood now ; but whenever I have lately chanced to pass the little street, I have seen the familiar, ugly form of 'Spot' sitting serenely, amid a group of children."

The authoress of " Lights and Shadows of Canine Life " thus describes a scene with her dog " Ugly " :—

" On going from the hotel at Linz to the station, in the omnibus, a man seated opposite to me noticed ' Ugly,' and warned me to be very careful in Vienna; telling me that every hour of the day carts were sent out, preceded by two men carrying nets full of hooks, which they threw over all dogs that were unmuzzled or loose. The dogs, when thus caught, are put into a cart, and carried to a place of confinement, where, if not claimed within twenty-four hours, they are killed.

" I asked many questions, of course ; and ' Ugly ' never took his eyes off the man, who presently observed how frightened he looked. I turned to him, and felt sure that the poor fellow perfectly understood what was said ; or at least knew something of the meaning of the word ' kill,' which he had heard before. When I had told him, in the journey, that they would kill him, he used to cry."

Captain Hall, in his description of the Esquimaux dogs, mentions one particularly " sharp fellow," whose name was " Barbekark " :—

" One day, in feeding the dogs, I called the whole of

them around me, and gave to each, in turn, a *capelin*—a small dried fish. To do this fairly, I used to make the dogs form in a circle, till each had received ten of these capelins.

"Now 'Barbekark' thought that he could play me a trick; so every time he received his fish he would back out of the circle, and thrust himself in again further on; so getting two turns every time I went round the circle.

"Each dog thankfully took his capelin as his turn came round; but 'Barbekark,' getting two turns, seemed to wag his tail twice as thankfully as the others. I thought I saw a twinkle in his eyes, when they met mine, which seemed to say, 'Keep my secret; I'm very hungry.'

"But seeing that I looked amused rather than angry, he was emboldened to go a step further, and to change his place twice in each round, so as to get *three* portions. This was too much; and I thought it quite time to reverse his game, by utterly thwarting him. Accordingly, every time I came to him, I passed him over: he got no fish; and though he changed his place three times, he got nothing. And if ever there was a picture of disappointment and of sorrow, it was to be seen in that dog's countenance, as he saw how I was punishing him. Finding that nothing he could do made any difference, he withdrew from the circle, and came straight to me, pushing his way between my legs; and looked up in my face, as if to say, ' I've been a very bad dog,—forgive me, and I will cheat my brother dogs no more.' So, as I went the round three times more, I let him have the fish, as he had asked for pardon and shown signs of repentance."

We must add a few more instances of canine intelligence. Mr. G. R. Pulman, writing to the *Naturalist's Magazine*, says,—

" A gentleman of my acquaintance, Mr. H——, of Axminster, was the owner of a very intelligent and sagacious bull-terrier of the largest size. It was singularly docile, and strongly attached to its master, of whom it was the constant companion. One day Mr. H—— had occasion to call at a house at the entrance into the town of Lynn Regis. He alighted from his gig, leaving his dog on the driving-box. Something frightened the horse, and made it start off at a tremendous pace towards the town, the reins trailing on the ground. In a few seconds the dog leaped from the gig, and seized the reins in its mouth, pulling them with all its strength, and allowed himself to be dragged for a considerable distance, till at last he actually succeeded in stopping the horse by pulling it round into a gateway. The dog still retained a tight hold of the reins—not relinquishing them till he saw that some person had hold of the horse's head."

Mr. St. John, in his " Tour in Sutherland," says :—

" There is a kind of quiet discretionary courage which some of these rough terriers have, which is very amusing. My dog 'Fred' is as much at home in a crowded railway station, or in a street, as he is in a furze-cover. In travelling, when he has seen me safely housed in an hotel, he soon wanders off in search of adventures of his own ; and though the town may be new to him, he invariably finds his way back to my room, spite of waiters, chambermaids, etc. I used to be afraid of losing him ; but after some experience, I found that it was best to leave him to his own devices. Once or twice I sent ostlers and boys in search of him ; but he always came back alone, looking rather ashamed, and not venturing to make himself prominent in the room till he had examined the expression of my face from under a chair or sofa; for dogs are great physiognomists. Then, on seeing that I am pleased at his return, he comes out of his place of refuge, wriggling his body in all sorts of coaxing attitudes, and at last putting his honest rough face in my hand or on my knee; glad that peace is declared, and waiting till he sees me hat in hand again."

" A sort of retriever, called ' Reves,' an old favourite of ours (writes a gentleman in Scotland), was in the habit of going for a walk before breakfast with my father.

One morning it so happened that my father did not intend to take his usual walk. 'Reves' soon became very impatient; and seeing no signs of his master, he got upon a chair in the hall, took his master's hat from the peg, carried it up to his room, and scratched at the

door for admission. As soon as the door was opened, in walked 'Reves,' laid the hat at his master's feet, and pushed his nose into his hand. The idea was entirely his own; he had never been taught to fetch a hat."

Mr. Wood tells us that—

"There was a dog at Margate, some years ago, who knew the use of money. He used to beg for pence, and take them to a baker in the High Street, to be exchanged for biscuits. One day the baker, curious to see how the dog would behave, took his penny and gave him a burnt biscuit. The next time the dog got a penny he took it to the baker, as usual, showed it him, and then went off with it to another baker, who lived nearly opposite. This he did again and again, showing the penny to the baker who had wronged him, and then transferring his custom to his rival.

An engineer wrote to Mr. Wood as follows :—

" I once lost a bet which I had made, that a little dog belonging to an ironmonger at Knighton would not take a penny off a red-hot bar of iron. The bar was heated red-hot, and no sooner was the penny laid on it, than the dog, without the least hesitation, dashed at it: the thing was done so quickly that I could not see how ; but the dog knocked the penny off the bar, and then sat down quietly, to wait till it got cool. His look of perfect self-satisfaction was most absurd. The penny went to the baker's, to buy a bun. The dog knew exactly the right-sized bun, and would keep his paw on the penny till he got it."

A fire broke out between two and three o'clock in the morning, in the house of Mr. R. Handley, in Bermondsey. Mr. Handley kept two small dogs in his house ; and that night, by some oversight, he had left his bedroom door ajar. Shortly before three o'clock, he and his wife were both awakened by the dogs scratching their faces. On getting up, they found volumes of smoke pouring up the staircase, so as to prevent their escape that way. Rousing the other inmates, they opened one of the back windows, leaped out upon some leads underneath, and escaped into the next house, from which they got into the street with safety. The fire was subdued in the course of an hour or two; but not until Mr. Handley's premises had been entirely destroyed.

"V. D." writes to *Land and Water* as follows :—

"A friend of mine, who would not have spared any pains in teaching his sheepdogs their proper lessons, had one dog who was not in the habit of doing his work in a satisfactory manner. In a country full of hills and valleys, it was often the duty of this dog to collect the flock at the bottom, and to drive them up the hill; instead of which he would drive them half-way, and then leave them to wander again. On the opposite side of one of these hills lived a farmer who had a dog of a more satisfactory kind, —both knowing its duty and performing it. On some occasions it was observed that, when the careless dog had failed in his task, the other would pass over from its own land, take charge of the flock, and drive them home. But its patience at last was exhausted;

and one day, when the flock had become dispersed, the watchful creature first finished the work,—drove the flock home,—and then fell upon the careless dog, which it severely punished. After this it refused to interfere further in the matter. It seemed to say, 'You rascal! what right have you to expect me to do your work?'"

In the *Science Gossip* we read:—

"Two ladies were leaving their house at Liverpool, to go into Cheshire. Just as all the furniture was packed up, and got into carts for removal, they were startled by a violent ringing of a bell connected with one of the bedrooms. Everybody ran upstairs to see what was the matter; when, on reaching the landing, there stood Master Gyp, a small spaniel, looking up at the bell, and wagging his tail with great delight. When the bell ceased to swing, he ran back into the bedroom, pulled the bellrope, and then came back to watch the bell ringing. He had only just discovered that the pulling of that rope set the bell ringing, and this discovery seemed greatly to delight him.

Mr. Swainson, in his "Zoological Illustrations," tells us that—

"The Rev. Mr. S——, living in Denbighshire, had

a favourite Newfoundland dog, who lived at large, partook of the best of everything, and exercised his power with great mildness. He was seen more than once to leap a gate which separated the yard of the house from the farmyard, in order to carry provision to another dog who was tied up in the stable."

In Bircham, in the county of Norfolk, a shepherd's dog was employed to take his master's dinner to him, while employed out in the fields. He carried it in a tin can with a lid; and on one occasion was met by another dog, who smelt the contents of the can. They had a fight, and the intruder was driven off. But in the scuffle the lid had been knocked off the can. The shepherd's dog did not know how to get it on again; but he saw that the two belonged to each other, so he adopted this expedient:—He carried the can in his mouth for some distance; then he returned and brought the lid, and carried it some way further. He then returned to the can, and carried that beyond the lid; and so at last he got them both safely to his master.

A correspondent of the *Animal World* says,—

"A day or two since, whilst walking on West Looe Quay, my attention was attracted to the actions of an

Irish water-spaniel which was coming towards me; when
suddenly it jumped into the river, where it is about one
hundred or more yards in width. On looking across, I
saw that a sailor on board a vessel on the other side was
calling the dog; and seeing it swimming towards the
vessel, he prepared to take it on board by means of a rope
which he threw over the side, holding an end in each
hand, thus forming a loop reaching to the surface of
water. The tide at the time was running in fast, so that
the dog was carried by it some considerable distance up,
and had to swim against the tide to reach the side of the
vessel; when it made an attempt to get its fore-legs over
the rope, but was unable to do so, the tide again carrying it up the river. A second attempt failed in the same
manner. Again the dog swam up to the side of the
vessel, and passed it without seeming to notice the rope,
and, as I thought, had given up any further attempt to be
taken on board; but no,—having swum six or eight yards
beyond the vessel against the tide, the dog deliberately
turns round and swims with the tide into the bight of
the rope, gets both fore-legs over, the rope crossing the
chest, the fore-legs pressed back so as to prevent its
slipping off the chest; and so the dog is drawn up the
side of the vessel, about ten or twelve feet in height,
and thus taken on board.

" Truly there is more than instinct displayed here. I
think that instinct is shown by the dog at first making

two attempts to get on board the nearest way, whilst swimming against the tide,—in which it fails; when reason steps in, and is shown by the dog taking the different course of swimming to the rope with the tide, and by that means accomplishing the desired end."

Another correspondent says,—

" There is a curly retriever at Arundel bearing the name of 'Shock,' which sets an example of good manners **and** intelligence to the animals which are **not** dumb. He carries the cat of the stables tenderly in his mouth, and would carry the kitten, but at present the kitten prefers **its own** means of locomotion. When Sangers' **elephant got into trouble in** the river Arun, this wise 'Shock' was sent **to turn him out; and** his perseverance succeeded. **He insisted on** carrying **a bundle of umbrellas** to the **station,** and safely delivered them **to** their owners; and then, with many wags of his brown tail, he demanded a halfpenny for his trouble. This halfpenny he carried to the nearest shop, laid it on the counter, and received his biscuit in return. Need we say this dog has a kind, sensible master ? "

Here is a third, from the same source :—

" There is in the village of Halton, near Lancaster, a Skye terrier, which has recently displayed a remarkable sense of natural affection towards its sister of a younger litter—the respective ages of the two being sixteen and eight months. Both being in the habit of going with the niece of the owner in her vocation of delivering letters in the village, they were accompanying her, as usual, to a cottage by the side of the river Lune. As they **were** indulging in their accustomed good-humoured frolics, 'Guess,' **the** younger of the two, had the misfortune to lose her footing, and rolled down **the sloping** embankment into the river. Great was the emotion of 'Fan,' as was evident by her piteous whine, and by her proceeding at once to the foot of the embankment, to ascertain what was the real position of her companion. **There she** found 'Guess,' who had partially raised herself **out of** the water, and was endeavouring with her forefeet to scramble up a low wall which forms a protection to the banks at the water's edge. 'Fan's' sagacity and presence of mind were here displayed, in trying to render all the assistance she could, by seizing hold of her companion's ear, which was within reach. Finding, however, that she was likely to lose her footing, and thus share the apparent fate of 'Guess,' she let go her hold, and poor 'Guess' had to undergo another ducking. But she soon managed to regain her former position; and 'Fan,' having taken the precaution to plant herself more securely, with her breast resting on the edge of the bank, again seized her little sister by the ear, and thus enabled her to scramble to the top of the wall, and rescued her from what might otherwise have been a watery grave."

The *Bristol Times* tells us,—

"Some little time ago a gentleman residing near Ascot let his house to a lady, who brought with her a number of dogs, including a little Dandie Dinmont terrier. The owner of the house and his wife had also left their pets at home,—among them a very fine Labrador dog. In a short time a warm friendship grew up between Labrador and Dandie Dinmont, which continued till the close of the tenancy of Dandie's mistress. When that event occurred, all the out-going lady's dogs were sent away together in a leash. By some means, however, Dandie loosened himself, ran away over the heath, and was lost. He did not return to the house he had left; and his mistress eventually mourned him as dead. Meanwhile, the lady to whom the Labrador dog belonged had returned to her home; and, after some days, was informed by her servants that there was something very odd about her favourite. He never ate but half his dinner, and carried off regularly his best bone,— disappearing each time for an hour. The lady ordered him to be closely watched and followed; and this being done, the 'good deed done in secret' by the generous animal came at last to light. He took his food, it seems, every day to his friend, poor little Dandie Dinmont, who had somehow got himself caught in a trap, and was lying lame and helpless on the heath—where he would, of course, have starved but for the charity of Labrador."

It is impossible to forget, in this place, the multitudinous cases upon record of the wondrous power possessed by dogs of going straight to a desired point, even over a distance of hundreds of miles. Here are two instances of this kind, which we borrow from the *Animal World :—*

"Our friend Mr. P——, of N—— vicarage, going up to spend some time near London, was in the habit of sending his groom and horses by the Stockton steamer; and on one occasion they were accompanied by a little terrier which lived habitually in the stables. On arriving at the London docks, the groom set out to ride through London, followed by the little dog, up to his arriving in Bond Street, where he missed it; and considering search hopeless, the groom and horses proceeded alone, giving up the dog as lost,—for he had never before been in London. The first letter from the north announced his safe return; he had simply trotted in and laid down in the stable. On inquiry at the docks, it appeared the dog had actually, after having retraced his way through the intricate streets of the city, picked out the Stockton steamer, which had moved to a different position in the meantime, from amongst the crowd of ships moored in

THE COMRADES.

the docks, made his way on board, gone direct to the place where the horses had been kept, and there waited the return voyage, and landed safely at Stockton."

"Last week Mr. Alex. Adams, of South Tyne Paper Mill, Warden, accompanied by the Rev. J. M. Wilson, left Hexham for a tour in the Holy Land; but previous to his departure Mr. Adams took his dog, a handsome black five-year-old retriever, to Edinburgh, and left it there with his sister, Miss Adams, who is at present residing there. On its arrival on Tuesday night it was fastened in the yard of the house, but during the night it broke loose and decamped. Nothing more was heard of the animal until Monday morning, when it turned up in an exhausted condition at South Tyne Paper Mill,—thus having been five days in travelling an unknown distance of upwards of a hundred miles. The dog has been once more despatched to Edinburgh by rail."

The Rev. James Hall, in his "Tour in Scotland," mentions a dog which was buried in one of the church-yards of Edinburgh, by a family of good position, who remembered a single instance of his attachment. His master had travelled from Edinburgh to Rome, and after remaining there for a time, left the dog with a friend, and returned home. Some time after his master's departure the dog felt a longing for home, and quitted Rome in search of his master. He passed over the Alps, and through France, till he reached Calais. Here he tried various vessels, but was driven away by the sailors; till at length a gentleman, who took a fancy to him, procured his admission to one of the Dover packets. This gentleman noticed him, paid him attention during the voyage, and thought he had gained his affection; but when they approached the Kentish coast, such was the creature's impatience that he leaped overboard, and swam ashore. In less than six weeks after leaving Rome he arrived at his master's house in Edinburgh, wasted almost to a skeleton by his journey, and by half-starvation.

Mr. Morris writes,—

"General W——, who kept hounds at Stanhope, in Durham, sent one of them to a friend and fellow-sportsman in Surrey. The dog was sent by sea from Sunderland to London, and from London it was conveyed to its quarters in Surrey.

"In due course, the General received a letter from his friend, thanking him for the present, which was much approved of; but adding that the dog, after being kept some time in the kennel, had been taken out hunting; that he had led the pack, etc., etc.; but that after the hunt was over he had entirely disappeared.

"About the time this letter reached General W——,

the servants at his house heard a scratch and a bark at the door, when one of them exclaimed, 'Why! that's so and so' (naming the dog). And, sure enough, on opening the door, there he was."

The chief fact in this case was that the dog had been carried to the south *by sea;* so that the road southward was wholly unknown to him. He had first to cross the Thames on leaving Surrey, and then to choose or find his way through Herts, Lincolnshire, and the midland and northern counties. How he could possibly do this is one of the mysteries of dog-life which no one has yet been able to fathom.

Mr. Moore, of Windsor, wrote to some friends in the north of England, asking them to obtain for him a well-trained greyhound, for one of the keepers of Windsor Great Park, for the purpose of killing fawns in the season. The dog was procured, and sent to town by the waggon. It arrived safely in Bishopsgate Street, from whence it was conveyed to the Belle Sauvage, in Ludgate Hill, and delivered to the driver of a Windsor caravan, who carried it to its place of destination in safety. The dog was kept in-doors, at Windsor, for two days, and received every possible attention from the family. After this it was left free; and in forty-eight hours it was nowhere to be found. A few days after, Mr. Moore received a letter from Yorkshire, informing him that the dog had reached its former home before the return of the waggon which had conveyed him to London.

A grazier in Lincolnshire sold a large lot of cattle to a London dealer, who took a fancy to a fine sheep-dog belonging to the grazier. On concluding the bargain, the seller agreed to make the buyer a present of the dog, if he wished to take him. The offer was gladly accepted; but as it was feared that the dog would not willingly leave his master, he was put into a covered waggon, closely shut up, and carried to London—a distance of more than a hundred miles. Here he was taken to a house belonging to the cattle-dealer, near Smithfield; and, lest he should make his escape, he was confined strictly for a fortnight,—in which time, it was hoped, he might become reconciled to his new home. But so soon as he **was** released, he quickly made **use of his liberty:** the cattle-dealer had to write **to his** former **owner** to **say** that the dog was **gone.** Both parties now looked upon the creature as **lost;** but before many days had elapsed, a piteous cry was heard at the grazier's door, and on opening it, the dog appeared, emaciated, footsore, and seemingly almost starved. How it could have **found its** way from Smithfield to its home in Lincolnshire,—a road which it had only once **passed** over, and then in a covered waggon,—is one of those mysteries which meet us so often when we examine the passage from place to place of these strangely-gifted beings.

Another correspondent of the *Animal World* writes,—

" ' Jack ' was a dog of no pretensions to beauty ; and he belonged to nobody. His coat was a dirty white one, with black and tan spots. He had been given when a puppy to some sailors, who threw him overboard, thinking him ugly and worthless. The fireman, however, had taken a fancy to him, and coming to the rescue, lowered a basket into the water and saved his life. From that time ' Jack ' always lived on board the steamers plying between Bristol, Swansea, and Cardiff. He became a great favourite among the sailors, and no passenger on board was more punctual to time. One day the stewardess of one of the steamers landed at the Mumbles in a boat, wishing to reach her home at Swansea sooner than she could have done had she waited for the tide which enabled the steamer to run into Swansea harbour. ' Jack' was not invited to go with her, but having some motive of his own for an early landing, he sprang after her into the boat. In some way, however, when the stewardess was getting into the omnibus which in those days carried passengers to and from Swansea, she missed him; and, greatly to her distress and mortification, she arrived at home finding she had left him behind. He had never, to her knowledge, landed at the Mumbles before, and she therefore feared he would not find his way. The steamer left Swansea and returned to Bristol on the following day, and still 'Jack' was missing. Soon after, however, a well-known scratching at the door, and the shouts of the children, announced his arrival. He was very tired and hungry, but rejoiced to find some of his best friends. After a good meal, and a great many pats and caresses from the children, he lost no time in setting off to the pier to watch for the steamer. The sailors had all missed him; and no sooner was the plank put across for the passengers to land, than 'Jack' sprang upon it, and bounded forward to receive the noisy welcome they gave him. At another time, when 'Jack' had been a passenger in one of the Cardiff steamers, he wandered up into the town, and so missed his return-passage to Bristol. Instead of being daunted, he walked to Newport, a distance of twelve miles, where he knew he should find a steamer going to Bristol; took his passage on board, and joined his own steamer at the Hot Wells. 'Jack' lived many years, and died of old age, much lamented."

The *Zoological Journal* describes to us some church-going dogs :—

" A poodle-dog belonging to a gentleman in Cheshire was in the habit of going to church, and remaining

quietly in the pew during service, not only when the
family were at home, but also when they were absent.
One Sunday a dam at the head of the lake gave way;
and the consequence was that the whole road was under
water. The congregation was naturally very small—
consisting merely of those who lived near to the church
—nobody from the manor-house being present. But
the clergyman afterwards informed the lady that while
reading the Psalms he saw his friend the poodle coming
slowly up the aisle, dripping with wet, he having swum
above a quarter of a mile to get to church. He went,
as usual, into his master's pew, and remained quietly
there until the end of the service."

Another church-going dog is reported from the west
of England. In a parish near Bath, the rector, years
ago, had a turnspit, of the most intelligent kind, who
followed his master about everywhere,—even into the
reading-desk on Sundays, to the no small amusement of
many of the congregation. Dr. B—— thought it was
time to put a stop to this, so he ordered " Toby " to be
locked up on Sunday morning. But this was to no
purpose; he scrambled somehow through the window,
and made his appearance, as usual, at the church.
Matters were now growing serious; so on the next
Saturday, when his work was done, the cook locked him
up in the wood-loft, from which there was no escape, by

window or otherwise.   So he merely made both day and
night hideous by his howls and cries to be let out.   Still,
it was hoped that at last he was conquered.   But another
Saturday came; and now, when noon was past, and
cook looked for "Toby," he was nowhere to be found.
Servants were despatched in all directions, but in vain.
Nothing could be heard of "Toby," till, on Sunday
morning, on entering the reading-desk, there Dr. B——
descried, curled up in a corner, the form and the twink-
ling eye of "Toby."   He had gained the victory, and
from that time forth was permitted to go to church.

We come, lastly, to the most interesting feature of all—
that of sympathy and personal attachment.

Mr. Hamerton says, in his "Chapters on Animals,"—

"We do not know the depth of the affection felt even
by the dogs we have always with us.   I had one who was
neither so intelligent nor so affectionate as others I have
known ; and, to my human ignorance, it seemed that he
did not love me very much.   But once, when I had been
away for weeks, his melancholy longing, of which he
had said nothing to anybody, burst out in a great passionate crisis.   He howled and clamoured for admission
into my dressing-room, pulled down my old things from
their pegs, dragged them into a corner, and flung himself upon them, wailing long and wildly where he lay,

till a superstitious fear came on all the house, like the forerunner of evil tidings. Who can tell what long broodings, unexpressed, had preceded this passionate outburst? Many a dark hour had he passed in silent desolation, wondering at that inexplicable absence, till at length the need for me became so urgent that he must touch some cloth that I had worn."

A correspondent of the *Animal World* writes,—

"My old friend Peggy Shorliker was a farmer's wife, of the old Lancashire type, who milked her cows, and took her own butter to market; and wore the short bed-gown and linsey petticoat. She had a great fondness for dogs; and 'Leo,' a young Newfoundland of mine, was a special favourite. He had a rather touching habit of planting himself erect in a sitting position before her; laying his chin on her knee, and gazing in her face as if he understood that she had some trouble, and as if he could feel for it; and she would stroke his ears and talk to him with all gravity, saying that he was 'as good at understanding as many a human; and was far better company than some.'

"'Where's "Leo?"'" was her first question, if by any accident the dog did not accompany me as usual. A piece of oatcake was always reached down for him, and the cat's supper of milk often went with it.

"Poor Peggy was afflicted with an incurable cancer;

and a consciousness of some evil, present or to come,
evidently pressed upon the dog's mind. One day, in
our joint visit, he delighted her by bringing in his
mouth a fine rabbit, which he carefully laid upon her
knee; and then, unable to express his feelings in
any other way, gave vent to them in a short, glad
bark.

"'He brought it on purpose for you, Peggy. A
weasel has just killed it in the plantation, and he
considers it will just be the right thing for your dinner
to-morrow.'

"'There's many a thing harder to believe than that;
for I'm welly certain there is a thought in this head for
his poor old friend,' said Peggy, as she stroked down his
ears in her wonted way, the dog all the time gazing
earnestly with his inquiring eyes into hers.

"The end was drawing near, when, in one of the last
days of the year, I called as usual in passing homewards. The poor sufferer was confined to her bed, in a
most distressing state. An artery had burst, and there
was blood about the bed, as well as in basins about the
room. I kept out the dog, thinking he might be in the
way. But 'Where's "Leo?"' was soon Peggy's inquiry;
'let him come; let him come just this once.' The
door was no sooner opened than the dog, who was close
to it outside, entered. He took no notice of any of the
disagreeables which were about; but walked gravely

up to the bed, raised himself on it by his two paws, and gently licked the poor face, half buried as it was in wrappings.

" 'Ah!' said Peggy, ' it's likely none of my own kind would have been so willing to kiss me as I am now.'

" There was a singular delicacy and tenderness in the manner of the dog as he thus gave her his last greeting, and then dropped quietly down by the side of the bed, —there sitting motionless, his head hanging down, solemn and sad. The scene was painfully touching. All were alike overcome. The daughter of the house and a stout servant lass both sobbed audibly.

" I shortly took my leave. The dog, instead of the usual gambols, crept quietly after me, and kept close by my side all the way home. He passed the usual pond without any of the usual expectation of a stick to be thrown in and fetched out; and on reaching home, he went at once straight to his kennel, with drooping head and tail waiting, to be chained up ; and so we parted for the night, as fellows in sorrow might do. It was the last time either of us saw poor Peggy. I had to leave home for a few days the next morning, and when I returned, she had entered into rest."

" My friend Mr. E—— lately called at a house," writes Mr. Morris, " where, as the master was absent, he had to sit down and chat with the mistress. After a time

a dog came into the room, whining and looking very
miserable. This was soon accounted for by the mistress
exclaiming, ' Oh dear ! Mr. —— is coming home tipsy;
and you will see that the poor dog, who is so fond of
him at other times, will not go near him now.'

" This was soon confirmed, for the master of the house
came into the room, and, after a few words, began to
coax his dog ; but the creature quite refused to go
near him. This, the mistress afterwards told me, was
invariably the case under similar circumstances ; the
dog would always shrink from his master's touch.

" One of the 17th Lancers confirmed this by a similar
instance. He said, ' Well, sir, there is a mare in our
regiment that is equally alive to these things. Her
rider will say, when he has taken more drink than is good
for him, " Now, if I don't mind, that mare will get me
into the guard-house to-night." ' The fact was, the mare
could tell when her master was sober, and when he was
not ; and when he had been indulging, she would
squeeze him against the wall, or in some other way
show her dislike to it ; and this generally gave rise to
a quarrel ; a row would ensue, and the tipsy man would
soon find himself in trouble."

Elihu Burritt tells us,—

" I was sitting at the breakfast table of a friend, who

is a druggist, when he was called into the shop by a neighbour who had come for medical advice and aid in a very affecting case.

"His family dog had incidentally made the acquaintance of a neighbour's child on the other side of the street. While lying on the doorstone, he had noticed this little thing, sometimes at the chamber window, and sometimes in a little carriage on the pavement below. During one of his walks he met the baby, and looked over the edge of the little carriage, straight into the eyes of the little child, and said, 'Good morning,' as plainly as he could.

"Day by day, and week by week, this friendship grew, increasing with the weeks and months of the little child. The dog, at last, having no children at home to make friends of, came to fix his thoughts mainly on this little child on the other side of the way; attending regularly the child's little carriage in its daily airings. Great was the delight with which he gave himself up to the pattings and rompings and rough handlings of those baby hands.

"But, one day, the baby, reaching out of the window above, lost its balance, and fell on the pavement below. It never breathed again. The poor dog saw its fall. His heart died within him, as he saw it taken up, quite dead. He uttered but one long whining moan of grief; but from that moment he could not eat. He refused to

be comforted—pining and crying day and night. He
would not stir from the spot where the baby's body
had lain. And now his master's errand to the druggist
was, ' Can you give me anything which will make the
dog eat his food ? ' "

"A clergyman," writes Mr. Morris, "who resided in
an out-of-the-way village on the banks of the Wye, had
a Scotch sheepdog, who, as the sequel shows, was **much**
attached to his master. On his death the dog was taken,
after the funeral (which he **had** followed to the grave),
to the **house of the** deceased's nephew, at about twelve
miles' distance. For some eight or ten days he seemed
tolerably contented. At the end of that time, one moon-
light night, he was missing. He must have crossed the
Wye; for he rushed into the kitchen of the parsonage
where he had lived so long, while the servants were at
breakfast. From the kitchen he rushed upstairs to the
door of his master's room, and from thence turned and
ran out. He was followed, and was found in the church-
yard, on his master's grave. He was busily employed
in tearing it up, and was covered with dust and earth.
He was taken back to the parsonage, and was kept
safely till the nephew sent for him. Being kindly
treated, he gradually became reconciled to his loss,
and lived contentedly in his new residence until his
death, which happened not very long ago."

The more recent case of "Greyfriars Bobby," of Edinburgh, obscures the following, but it also confirms it, by showing its probability.

Mr. St. John, in his " Tour in Sutherland," says,—

" A minister of a parish in this neighbourhood having died, his favourite dog followed his body to the grave, and no inducement could persuade the faithful animal to leave the place. The people around, finding all their endeavours to draw him from the grave quite fruitless, and honouring his fidelity, brought him food and protection from the weather. This went on for several weeks, until a new minister came to the manse, and his wife, from some wretched kind of superstition, ordered the dog to be killed. The source," adds Mr. St. John, " from which I received this narrative, leaves no doubt on my mind as to the certainty of the occurrence."

" About twenty years ago," says Mr. Morris, " there lived in a little hut beneath the Southdown Hills an old man who had been a shepherd in that country for more than sixty years. He was poor and badly off—a small weekly allowance from the parish being all he had to support him. His only companion was an old dog, whom he called ' Bob.' Poor ' Bob,' like his master, lived a very hard life; being, in fact, half-starved. One day a gentleman who was passing the hut, and who lived in

the next village, persuaded the shepherd to let him take
'Bob' home with him.   In his new home he had as much
as he could eat, a comfortable bed every night, and the
greatest kindness every day.   But 'Bob' was not happy.
After a fortnight's stay, the gentleman thought he
might take him out for a walk ; but no sooner had 'Bob'
got outside the door, than he scampered off as fast as
his legs could carry him, back to his old master and his
miserable home.

"Six months after this, poor 'Bob' was found one
morning standing in the gentleman's garden, looking
very sorrowful and very thin.   On the door being opened,
he walked in, laid himself down, and stayed contentedly
all night.   The next morning he was taken back to the
hut ; and then it was discovered that the poor old shepherd was dead, and had been buried.   Very willingly
the gentleman took poor 'Bob' home with him, and
treated him very kindly, till the faithful creature died
of old age."

The *Portland Press* (U. S.) vouches for the following
story :—

"A short time ago a female Newfoundland dog
made the acquaintance of a lady of the city, who would
often throw it a piece of meat, or a bone; until, at
length, it became quite a matter of course, every day,
for the Newfoundland to appear, and for the lady to

have something ready for her. The appearance of the animal changed, and the lady began to understand that there **must be** puppies in the creature's home. So one **day she** said, while feeding her, ' Why don't you bring me one of your puppies ? ' repeating the question several times, **the** creature looking in **her** face with **evident** intelligence. The next day, to the lady's astonishment, the Newfoundland came, bringing with **her a** little puppy. The lady fed **them both,** and then took up the puppy into the window, at which the **mother** seemed well satisfied, and quitted the place quietly, not appearing again for three days. Then she came back again, and after feeding her, the lady said, ' Next time bring all your puppies; **I want** to see them.' Sure enough, the following **day, the mother returned,** bringing with her three little puppies. Several of the **neigh-** bours witnessed the whole transaction, and declared that it was about the most wonderful proof of dog-sagacity that they had ever witnessed.''

Mr. **Lane** writes to Mr. Morris,—

" In the **hurry of** going on **board an Irish** packet boat we left a small wire-haired terrier in the hotel sitting-room. Not being called, he did not show himself. We did not discover his loss until the boat was on its way. On reaching Cork a halfpenny was well breathed upon, and enclosed in a letter to the landlord, with directions that if the dog had not already allowed himself to be caught, the coin was to be given to him. We heard afterwards that when the servants entered the room he retreated under the sofa, and utterly refused to come forth, either for bones or for fear of broomsticks. But so soon as the letter was received, the halfpenny was thrown to him, and he came out at once with it in his mouth. He then allowed himself to be put into a basket, tied down, and sent on board the packet without a single struggle, or even a growl."

A couple who lived on the side of the Ermerdale Hills, in Cumberland, often took their little girl with them when they went into a neighbouring wood to gather fuel. One evening, in searching for wild flowers, she strayed out of sight; and when twilight and darkness came on, they searched for her in vain. At last they went back to their cottage, in the hope that the child might have wandered back. Finding that she was not there, they got torches and renewed the search, but still without success. Tired out, they went home again, and the mother mechanically spread the supper-table, when their dog jumped up, seized a lump of bread, and rushed out of the cottage. The father said, " I never knew the dog to steal before." Before daylight the search was renewed, but still in vain. When breakfast-time came, the dog appeared again, and repeated his extraordinary conduct of the evening before. The wife, struck with a thought which was full of hope, exclaimed,

" I'm sure he knows where the child is ! " Instantly they both started forth and managed to trace the dog, and found him on the edge of the lake, and the child

holding in her hands the bread which he had just brought her.

" It is generally believed," says a correspondent of the *Animal World,* "that bulldogs are very stupid and savage animals ; but it is not always so. What I am about to relate took place at Eastbourne, Sussex. A gentleman of my acquaintance possessed a very fine but fierce bulldog. When first I saw him he came up to me and kept smelling me for some time, and wagged his tail as if highly pleased. Some little time afterwards he got one of his legs broken, and the owner asked me whether I would set it. Well, I must own I did not much like the job, for if he would not lie still, there was no holding him still on account of his great strength ; but I consented to do it. I patted him, and then let him lick my hand ; we then put him on the sofa, where he lay all the time. When I pulled the bone out to put it in its proper place he closed his eyes with the pain, but did not otherwise move. When I went to see how the leg was getting on he would lie on his back, and put up his leg for me to look at, and was not satisfied unless I fingered it a little. It began to get well speedily ; but he broke it again, and underwent the operation with the same passiveness, though it must have been more painful than before. The leg began to mend again, slowly but surely, and is now quite well. But though he was so quiet and gentle with his master and myself, he would not let any one else touch his disabled limb. I have written this, trying to defend bulldogs from the cruel floggings they are subjected to, upon the false principle that they are more brutal and savage than other dogs, and ought to be treated in a more brutal manner."

A lady writes,—

"Last year I spent a month in a German town, and was much distressed to see a poor watchdog tethered all the day long in the broiling sun, near the Kırsaal. Whenever I was able to go out I took him bread in my pocket, and changed his dirty water for him; but to get him a run was impossible. He was chained to one spot at the beginning of each season, and kept there till it was over. Some months later in the year I drove over from my residence to spend a day at the said town. The Kursaal was closed, and I never thought of my little friend; but while I was walking about in the gardens with a lady, all at once a great dash was made at me, with expressions of gratitude too vehement to be mistaken. The poor dog was now at liberty, bounding with joy to see me. My hands must be licked again and again, and he must roll over and over at my feet; and then nothing could induce him to leave us till he had escorted us safely to our apartments."

In the *Animal World* we read,—

"A clergyman residing at Sunderland, a small fishing village near Lancaster, had been dining with some friends about three miles off. He left their house, accompanied by a small dog, to return home between ten and eleven o'clock p.m. For the purpose of saving time and distance he took the sands of the seashore, the tide being out at the time; observing for his guide a lighthouse situated at the mouth of the river Lune. By taking this course he would save about a mile. The night was dark. He walked on and on for a considerable time, but was surprised in not finding himself sooner at his destination. The painful truth of having been misguided, and that he had therefore missed his way, now flashed on his mind, as he found himself being surrounded by the flowing tide. He knew not what to do. His danger became imminent. He shouted for help and guidance; but he was beyond all human hearing. His situation was a perilous one. He saw no way of escape. Death stared him in the face. He therefore gave himself up for lost. He then knelt down, and committed himself to the care of his Maker, expecting that a few brief moments would terminate his earthly existence. Just at this fearful moment his faithful dog, which in his perplexity he had forgotten, rubbed against his leg. This brought hope and relief to his mind. 'What, Jock, is that you?' asked the clergyman. The

dog, as if to assure his master that it really was he,
wagged his tail, and looking earnestly into his face,
whined very piteously, then ran a short way from him,
as much as to say, 'You are in danger; but follow me,
and you will soon be safe!' The clergyman did so—
now through deep water, the dog swimming by his side
or on before him, then crossing deep gullies, immersing
him up to the middle in water. Thus, following his
guide, he ultimately found himself in a place of safety;
having, by the sagacity, fidelity, and intelligence of his
dog, escaped an untimely end. The light of the lighthouse the clergyman had taken for his guide had been
extinguished at an early part of the evening, and another
lighthouse lighted further out to sea. Hence the mis-
take, and the danger incurred."

The *Dundee Advertiser* publishes the following:—

" A striking example of that devotedness and faith-
fulness characteristic of several species of the canine
race to their masters, may be witnessed at the door of
the Dundee prison leading from the police-office. At
the police-court, the other day, John Melville, a shepherd, was sent to jail for seven days for drunkenness.
The shepherd possessed a beautiful collie, which
patiently waited upon its master during his trial. On
leaving the bar and being marched to prison, the faithful
animal followed, and would have willingly shared a

corner of his cell had it been permitted. The good
dog, being necessarily separated from its master, would
not, however, desert his place of confinement, but took
up a position at the prison-door, where it still keeps
'watch o'er his lonely cell.' Meat and drink were laid
down to it by one of the police officials, but when, some
time after, another supply was brought, it was found
that the poor brute had not even tasted the first, and
no coaxing could induce it to do so; neither can it be
induced to accept a warmer and more comfortable place
of rest."

Prebendary Hey, of Lichfield, tells us,—

"A bull-terrier at Clay Cross, Derbyshire, was re-
joicing over a litter of pups, attending to them with
motherly care, caressing them, and showing for them
much anxiety. The owner of this interesting family,
accompanied by a few friends, came to look at them.
The fierceness of the mother was instantly developed,
and was manifested by impetuous growlings, snarls, and
barkings. She would evidently die in defence of her
offspring. But her master said, 'I can take every one
of those pups from the mother, and she shall not hurt
me.' He drew near, and approached his hand towards
the pups. The mother flew at it, and seized it in her
mouth. He did not attempt to withdraw it until she
released it. He proceeded to lay hold of one of the

pups. She seized his hand again and held it fast, but without biting it. By degrees he withdrew one of the little creatures. The same process was repeated until every pup was removed, and the mother was bereft of all her offspring. She was the picture of misery. She looked up piteously in her master's face, and howled with an exceeding bitter cry. The appeal could not be resisted. He replaced her little ones in their nest, and rejoiced to witness the caresses and gratification of the mother. But his surprise and that of his friends was great when he saw her take them up one by one in her mouth, and bring them and lay them down at his feet. He waited to see what she meant, and was soon convinced that she wished to express her entire confidence in her master. She laid her dear ones at his feet, one by one, with an assurance that he would take care of them with an affection equal to, if not greater than, her own. She then returned to her nest, now empty, lifted up her head into the air, and gave vent to several piteous cries, until her little ones were once more restored to her. It is impossible to misunderstand her impulse. She at first had misdoubted her master, but now she trusted him. She brought them of her own accord and placed them at his disposal, with full confidence in his love.''

A lady writes to *Land and Water*,—

4

"A spaniel which we had for some years, and which was rather addicted to hurting cats, took a fancy to a young kitten, to which he became so much attached, that he allowed it to eat off his plate, and to sleep in his kennel, and was seldom seen without it, though none of the other cats dared to approach him. In time the kitten grew into a very fine cat, but the affection of the two seemed to increase rather than diminish. Puss would watch for the spaniel's return from his walks ; and when neither were otherwise employed, she would take a nap between his paws. Sometimes the dog, who was of a pugnacious disposition, would come in bleeding from an encounter with some neighbouring dog, and pussy would run to him with every sign of sympathy, licking his wounds and comforting him in every possible way. At last, the spaniel being ill, they were separated for about three weeks; and the cat's delight when he was restored to liberty knew no bounds. But, soon after, the dog died, almost suddenly, while absent from the cat. Her grief was intense. For several weeks she wandered about the stable-yard, where he was buried, refusing all food, till she was reduced to a skeleton. We were obliged then to use force, and to pour milk down her throat with a spoon. She continued in this miserable state for some time, till, happily, a kitten of another breed attracted her notice, and she began to show signs of fondness for it. From this time forward she visibly improved, until, at last, she regained her former comely appearance."

The contractors engaged on the Boston waterworks (U. S.) had a valuable cart-horse injured some time ago. The animal was led home to the stable, where a great number of horses were kept. The ostler had a water-spaniel, which had been for some time always about the horses, living in great intimacy with them. When the disabled horse was led in he lay down, and began to show signs of pain and distress. The dog at once ran to him, and began to lick his face, and in various ways to manifest his sympathy with the sufferer. He sought the master, and drew him towards the horse, and showed great satisfaction when his master bathed the wounds, and in other ways ministered to the animal's wants. At night the ostler told the dog to go to bed ; but he refused, and preferred to remain with the horse all night. Forty-eight hours after the accident had occurred, the dog was still in the stable, and it was believed that he had scarcely slept. He allowed no one to come near the horse except the master and those who belonged to the stable. His whole appearance was one of extreme distress and anxiety. He often laid his head on the horse's neck, licking his face; and the horse evidently felt his sympathy and kindness.

A district registrar of Edinburgh was in the habit of

spending his holidays with a friend who had a sheep-
farm on the Ochill Hills; and in these visits he had
formed an acquaintance with one of his host's dogs.   In
the autumn of 1859 he was with his Stirlingshire friends,
and inquired particularly for his canine friend; when
the gudewife told him that " he was deed an' gane."
Mr. H—— remarked that his death must be a great loss,
as he was such a wise animal.   " Indeed, ye may say
that," said the gudewife; " we hae had mony gude
tykes sin' we came to the farm ; but we never had one
like him.   After ye left us last year, 'Rab' left hame one
afternoon, and about the gloaming he brought a sheep
down from the hill ; and, would ye believe it ? he gaed
nine times to the hill that evening, and brought a sheep
hame every time ! "   " But," asked Mr. H——, "why did
he bring the sheep home ? "   " That's just it," said the
gudewife; " every one of those sheep was diseased,
and he brought them hame, both to be cured, and to
prevent the rest of the flock from catching the disease."

Of the unspoken language, the good understanding
that often exists among animals, we have many proofs.

" Two of my friend's dogs," writes a London phy-
sician, " had a special attachment to and understanding
with each other.   The one was a Scotch terrier, gentle,
and ready to fraternize with all honest comers.   The other

looked like a cross between a mastiff and a large rough staghound. He was fierce; and it was wise to cultivate his acquaintance before you took any liberties with him. The little dog was gay and lively, the other was stern and thoughtful. These two dogs were often observed to go to a certain point together, and then the small one remained behind, at a corner of a large field, while the large dog took a round by the side of the field, which ran up hill for nearly a mile, and led to a wood on the left. Game abounded in these districts, and the nature of the plan of the two dogs was soon perceived. The terrier would start a hare and chase it up the hill towards the wood, where they would arrive somewhat tired. Here the large dog, fresh and in wind, darted after the animal, who was easily captured. The two dogs then ate the hare between them, and quietly returned home. This plan of operations had been going on for some time before it was fully understood."

"A gentleman in Dumfries-shire had a dog and cat which were much attached to each other, and both were great favourites in the household. The dog, however, was not meant to sleep in the house, but was carefully turned out into the yard every evening; and yet, strange to say, he was always found in the morning lying before the fire, with the cat by his side. One evening the master of the dog heard a sort of rap at a back-door leading to the kitchen, and saw the **cat spring** up and **strike** the latch, while the dog pushed open the door, **and entered** without let or hindrance."

A correspondent of the *Animal World* writes,—

"**A** few years ago **a** farmer, **from** near Dumfries, walked to Penrith with his sheep **and two** dogs. Having sold his cattle, he prepared **to return** home; but 'Fan,' in the meantime, had **had puppies,** and was left in charge of a friend, **who made a bed for** her in his parlour, and **fed her.** After **a few** weeks, on **coming as usual** to feed her, neither she nor her puppies **were to be seen.** He looked for his hat: it had also disappeared! 'Ah!' thought he, 'the thief has taken that too.' After a diligent search, that proved useless —for neither dog, puppies, nor hat had been seen or heard of—the friend wrote to the Scotch farmer to inform him of his loss. **A few days after** he received the **following** reply: '*Make no more researches.* " *Fan* " *arrived here early this morning, with her three puppies in your hat!*' **Mark the** animal's *reasoning,* we may call it. She had seen the Penrith farmer put his *head* in the *hat;* 'Fan' seized the idea, and placed her little darlings in it—being no longer able to remain away **from** her master, or to **leave** her young behind."

A spoilt and petted little dog, named "Beau," and

his mistress, were on the sands at Penmaenmawr, in North Wales. The tide overtook them, and they were cut off 'from the beach. The lady was soon rescued by a bathing-machine; but poor foolish " Beau " was afraid to enter it, and remained on his little bit of sand, which was rapidly growing less and less. The lady writes,—

" When I found myself on the beach, I looked for my dog, thinking that he would probably swim after the machine. But no: the little idiot was still on the slip of sand, yelping and barking in great distress. I called him, bidding him swim across, as I knew that he could use his limbs in the water as well as on the land. But he would not come, and the sea was rapidly rising, and poor 'Beau' had scarcely space whereon to stand and whine.

" Playing near me on the beach was a large, rough-haired dog—a sort of retriever. He soon saw the trouble we were in, and suddenly dashed into the water, and went up to 'Beau' and said something to him. I don't know what he said, but I have no doubt that he urged 'Beau' to swim across to his mistress. But alas! the brave dog returned to land again, but no 'Beau' with him. And now the sea was rising round my little terrier, and he himself was like a black-and-tan tiny island. The brave retriever dashed for a second time through the water, and stood beside poor shivering, yelping 'Beau.' Then he went behind 'Beau,' and very gently but firmly pushed, pushed, pushed him through the water towards the place where I was standing. As soon as they were both fairly in the deep sea, and it seemed a case of sink or swim with Master 'Beau,' the brave dog let him go, and with a few vigorous strokes brought himself to shore. And soon 'Beau,' now that there was no help for it, exerted himself, and quickly presented himself dripping and breathless at my feet. The brown dog, like a hero, made no fuss about what he had done; and I had nothing to give him but a pat on the head. His master was not on the beach, and I do not think that I ever saw the dog again."

The inundations which have recently afflicted France, and other parts of the continent of Europe, have repeatedly called into exercise the great power, in the water, of Newfoundland and other water-dogs. The following story, however, dates rather further back :—

"A very touching incident," says a Swiss paper, "took place on the bank of the river Seine on the 18th instant. Mr. L——, the proprietor of a floating bath, in which he lives, was breakfasting with his family and one of his friends. His attention was suddenly attracted by the fall of a body into the water. It was that of an unfortunate lunatic, who, having escaped from the hands of his

THE INUNDATION.

keepers, had just thrown himself into the Seine. Mr.
L——, who was accustomed to swim, plunged in to the
help of the unfortunate man, careless of the risk which
he ran in doing so, and very soon reached him. Then
occurred a scene at once strange and terrible. The
unfortunate lunatic tightly grasped his deliverer, who
struggled with him in vain, and both were on the point
of disappearing. The friend of Mr. L——, seeing the
peril which menaced him, leapt into the water to his
assistance. He was seized in his turn by the unfortunate maniac, and at the same time by Mr. L——, who
had almost lost consciousness; and all three were on
the point of perishing, when an unexpected deliverer
appeared, in an enormous dog, who had up to this time
watched the scene from the banks. The animal, throwing himself into the water, reached the struggling group,
and seizing the madman by the head, brought him to
land. Mr. L—— and his friend, thus delivered from his
grasp, were able to regain the bank. It is needless
to say that the dog was made the subject of a perfect
ovation."

Mr. Jesse, in his " Gleanings," tells us of a dog of the
Newfoundland breed, who had come to understand the
use of the pump, and how to avail himself of it. When
he was thirsty, he would go to the kitchen, take up a
pail, and carry it to the pump. There he would quietly

wait till one of the servants came by, to whom he in-
dicated, by gestures, that he wanted the pail filled.   And

then, having satisfied his thirst, he would quietly carry
the pail back to the place from which he had taken it.

In dealing with our home favourites, it is natural, after the dog, to turn to the cat. We cannot place them on a level; still, there are some respects in which pussy may even claim a preference.

One of Mr. Wood's friends wrote to him as follows:—

"A cat of ours once showed great self-denial. She was a terrible eater of small birds, chickens, etc.; and therefore, when on one occasion she was found to have passed the night in our aviary of doves, great was the alarm. But on inspection, not one dove was found to be missing; and though she was asleep in an inner cage, close to a nest of young doves, she had not touched a feather. And this was the more remarkable, in that, when she was released, she ate ravenously."

A correspondent of the *Animal World* writes,—

" My cat is in the constant habit of opening the back-door of my house, which has a thumb-latch. Almost immediately after she was brought to this house, eight years ago, she adopted this plan of letting herself in, to the no small amazement and often alarm of strange servants. When it is desired to keep her out the door has to be bolted; and at one time when she found this the case she used to go to the scullery-window, jump up

at that, and hang on to the upper part until her weight
pulled it down far enough to admit her.   The window
sash at that time ran down easily ; since new cords have
been put in she cannot do it : but the practice showed
something akin to reason, I think.   Another peculiarity
she has, which I never noticed in a cat.   She will not
answer to the call of ' Puss, puss ! ' but let me be where
I may, if I snap my fingers she will come directly.   I
have often tried her when she has been playing in the
garden and I have been at one of the top windows ; she
will immediately run in and come upstairs to find me.
But like a dog answering to his master's whistle, she
will not answer any one else.   The servants have repeatedly tried to imitate me, but she does not notice
their call.   I am convinced that cats are quite as intelligent as dogs, if their faculties are equally drawn out.
I once had a beautiful cat, whose kittens, when she had
any, I always used to go and look at every day—the
mother being very proud of the attention.   Once when
she had one, I happened to be very ill and confined to
my room.   No doubt pussy missed the usual attention ;
but she was determined her child should not remain
unknown to me.   The very first day I was able to come
down to the drawing-room sofa, pussy brought her kitten
in her mouth from the back-kitchen, where its bed was,
and laid it down at the kitchen-door, while she asked,
as plainly as possible, to have that door opened for

her; when this was done she carried the kitten to the drawing-room door, returned to the kitchen, and asked again for the other door to be opened; then brought it to my feet, laid it down in triumph, and looked at me. This process was repeated every day till I was able to go about the house; and though the kitten was generally sent back to its bed very soon, she never brought it a second time on the same day; merely, I suppose, thinking she ought to let me see it daily, if I could not go to it."

A family residing at Newcastle-on-Tyne went one summer to Tynemouth, leaving their house in the care of two female servants. One evening, when the servants were sitting together in the kitchen, their attention was attracted by a cat, which went up into a laundry over the kitchen, and then returned to them and mewed. She did this repeatedly, till at length the servants began to wonder at it, and went upstairs to see if they could discover the cause. When they went into the laundry, and looked about, they discovered a man trying to conceal himself in the chimney. One of the girls fainted away; but the other screamed out, and gave an alarm to the neighbours. Meanwhile, however, the man had escaped out of the window, and over the roof of the adjoining houses.

The attachment which cats are often found to form

for houses in which they **have lived** is a remarkable feature of their character.   Mr. Jesse tells **us of** one cat which a lady at Glasgow received from a friend at Edinburgh.   It was sent carefully secured in a closed basket, and for two months it was constantly watched. **But** having then given birth to two kittens, the watch-**fulness** ceased, it being supposed that she would now remain quietly by her offspring.   Soon, however, she disappeared, and the kittens with her.   In about a fort-night her mew was heard at the door of her old home in Edinburgh, and with her both the kittens.   The dis-tance between the two towns is forty miles.   She had **not** only to discover her road, but to convey her kittens, **and,** it is concluded, one at a time, returning for the **other.**   In this way she would have to travel one hun-dred and twenty miles.

Mr. Chambers, in his " Miscellany " (vol. vi.), gives an account, sent **to him** by **a correspondent, of a cat** which showed more **than ordinary attachment** :--

" A cat with her kittens was found in a hole in the wall, in the garden of the house where we lived.   One of the kittens, a beautiful black one, was brought into the house, and soon attached himself especially to me. I was in mourning at the time, and perhaps the simi-**larity of the hue of my** dress to his sable fur might have **attracted** him ; **but,** whatever **was the** cause, he never came into the room without immediately jumping into my lap, and purring and rubbing his head against me in a very coaxing and endearing manner. I went to Dublin every winter, and in the summer into the country; but the change of abode never seemed to trouble my favourite, so that he could be with me. Often, when we have spent the day out, he has come running down the street to meet us, showing the greatest delight. Once, when I had an illness, poor " Lee Boo " deserted the parlour altogether, though he was caressed by every one there. He would sit for hours disconsolate at my door, till he could steal in and jump on the bed, showing his joy by loud purring and by licking my hand. But as soon as I went down he returned to his regular attendance in the parlour."

Another correspondent of Mr. Chambers writes :—

" We have at present a cat who has formed a very warm friendship with a large Newfoundland dog. She is constantly caressing him, runs in all haste to meet him when he comes in, rubs her head against him, and then purrs delightfully. When he lies before the kitchen fire she uses him as a bed, pulling up and settling his hair with her claws, to make it comfortable. As soon as she has arranged it to her liking, she lies down and composes herself to sleep—generally purring till she is no longer awake ; and they often lie thus for an hour at a time.

Poor 'Wallace' bears the rough combing of his locks with patient placidity, turning his head towards her, and giving her a benevolent look, or gently licking her."

A little black spaniel had five puppies, which were considered too many for her to bring up. As, however, the breed was highly valued, her mistress was unwilling to destroy any of them; and she asked the cook whether it would be practicable to bring some of them up by hand. The cook observed that the cat had just kittened, and that, perhaps, some of the puppies might be substituted for the kittens. This was tried: one by one the kittens were got away from their mother, and two of the puppies were substituted for them. And here it was remarkable that the two puppies nursed by the cat were, in a fortnight, as active and playful as kittens usually are at that age; they had the use of their legs, barked, and gambolled about; while the other three, nursed by their mother, were whining and rolling about like fat slugs. The cat gave her two puppies her tail to play with, and they were always in motion; they soon ate meat; and, long before the others, they were fit to be removed. This was done, and puss became inconsolable. She wandered about the house, looking for her pets. At last she fell in with the spaniel, the mother, who was nursing the other three. A quarrel arose, ending in a fight, in which the spaniel was worsted, and puss walked off with one of the three. A day or two after she returned to the charge, and carried off another. But beyond the two—the number of those first entrusted to her—she did not go. The spaniel was allowed to keep one.

Of the wonderful intuitive faculty by the aid of which cats, as well as dogs, can trace their masters, Mr. Wood gives the following instance:—

"The cat in question was the property of a newly-married couple, who resided towards the north of Scotland, where the country narrows considerably by reason of the deeply-cut inlets of the surrounding sea. Their cottage lay near the coast, and there they remained for some months. After a while they changed their place of abode, and took up their residence in a house near the opposite coast. As the intervening country was so hilly and rugged that there would have been much difficulty in transporting the household by land, a ship was engaged; and, after giving their cat to a neighbour, the master and mistress proceeded by sea to their new home. After they had been settled there some weeks, they were surprised by the sudden appearance of their cat, which presented itself at their door, weary, ragged, and half-starved. As may be supposed, she was joyfully received, and soon recovered her good looks. It is scarcely possible to conceive how the creature could have been

guided.   Even had the travellers gone **by** land, it
would have been astonishing if the cat, days afterward,
should have been able to trace the line of journey.   But
in this case the travellers went by sea; yet the cat,
going **by** land, discovered them, and went at once to
their new dwelling.   The guiding faculty must be
something which passes our comprehension."

"There were," says Mr. Morris, "a few years ago, in
the family of a friend of mine, two cats, one known by
the common name of 'Tom,' the other by the equally
familiar cognomen of 'Pussy.'   'Tom' was much the
elder of the two, but for years he had lived in conjugal
harmony with 'Pussy,' who had been to him a faithful
and affectionate partner.   In peace and contentment
they shared the kitchen-hearth, regarded by all the
servants with respect and good-will.   But, as time wore
on, poor old 'Tom' began to show signs of wearing out:
decrepit, rheumatic, and stiff, with a coat that seemed
moth-eaten; while life was evidently becoming a burden.
At last the order was given to John, the gardener,
that poor old 'Tom' should be shot.   The order was
obeyed, and in a few hours poor old 'Tom' was dead and
buried.   No one knew that 'Pussy' had witnessed her
husband's death; but it soon became evident that she
must have done so.   She had been accustomed at meal-
times to sit on John's lap,—'Tom' preferring the society

of the cook; but from this time forward she never would go near John. She refused all food, and all caresses; until, after several days' fasting, she was induced by hunger to eat one small mouse. She sat, day after day, in all weathers, on the grave of her lost partner. For a whole fortnight she pined and wasted away. At the end of that time she disappeared. Her body was never found; but it was evident that she had sought some solitude in which to die."

A cat in a Swiss cottage had taken poison, and came in a pitiful state of pain to seek its mistress's help. The fever and heat were so great that it dipped its own paws into a pan of water,—a most unheard-of proceeding in a cat. She wrapped it in wet linen, fed it with gruel, and doctored it all the day and the night after. It revived: and now it could not find ways enough to show its gratitude. One evening, after she had gone upstairs to bed, a mew at the window roused her. She got up and opened it, and found the cat, which had climbed up a pear-tree nailed against the house, with a mouse in its mouth; this it laid as an offering at its mistress's feet, and then went away. These tributes of gratitude were often repeated from time to time.

Mr. Buckland tells us of a cat who had been accus-

tomed to the water, and who lived at Portsmouth.
His master, a fisherman of Portsmouth, thus described
him :—

"He was the wonderfullest water-cat as ever came
out of Portsmouth harbour; and he used to go out a-
fishing with me every night. On cold nights he would
sit in my lap while I was fishing, and poke his head out
now and then ; or else I would wrap him in a sail, and
make him lay quiet. He'd lay down on me when I was
asleep; and if anybody come he'd swear a good one,
and have the face off of them if they went to touch me :
and he'd never touch a fish, not even a little teeny pout,
if you did not give it him. I was obligated to take him
out fishing, for else he would stand and youl and marr
till I went back and catched him by the poll and shied
him into the boat; and then he was quite happy. When
it was fine he used to stick up at the bows of the boat,
and sit a-watching the dogs (the dog-fish). The dogs
used to come alongside by thousands at a time, and
when they was thick all about, he would dive in and
fetch them out, jammed in his mouth as fast as may
be, just as if they was a parcel of rats ; and he did not
tremble with the cold half as much as a Newfoundland
dog : he was used to it. I learnt him the water myself.
One day, when he was a kitten, I took him down to the
sea to wash the fleas out of him ; and in a week he
could swim after a feather or a cork."

"A shoemaker in Edinburgh," says Mr. Morris, "happened to leave the door of a lark's cage open, of which the bird took advantage, and escaped. About an hour afterwards, a cat belonging to the same person made its appearance, with the lark in its mouth. It contrived to hold it by the wings, so as not to cause the bird any injury. Dropping the lark upon the floor at its master's feet, the cat mewed, and looked up into his face, as if expecting his approval."

Mr. Hamerton says,—

"One evening, before dinner-time, I had occasion to go into a dining-room where the cloth was already laid, the glasses all in their places, and the lamp and candles lighted. A favourite cat, finding the door ajar, entered softly after me, and began to make a little exploration. The first thing he did was to jump upon a chair, and thence upon the sideboard. There was a good deal of plate and glass upon that piece of furniture, but nothing which, in a cat's opinion, was worth purloining; so he brought his paws together, and sat for a minute or two, contemplating the long glittering vista of the table. As yet there was not an atom of anything eatable on it; but the cat probably thought he might as well ascertain whether or not this were so by a closer inspection, so at a single spring he cleared the space between, and alighted noiselessly on the table-cloth. He walked all over it, and left **no** trace; he passed among the slender glasses, disturbing nothing, breaking nothing anywhere. When his inspection was over he slipped out of sight, having been perfectly inaudible from the beginning.

"One day I watched a young cat playing with **a** daffodil. She sat on her hind-legs and patted the flower with her paws, first with one paw and then with the other, making the light yellow bell sway from side to side, **yet** not injuring **a** petal or a stamen."

Mr. Wood had **a** favourite **cat** which he had brought up from infancy, and **which** was on terms of the greatest familiarity with him. One day puss came to the door of **the** dining-room, mewing piteously. Mr. Wood opened it, when the cat went to the foot of the stairs, ran up two or three of them, and looked round for Mr. Wood to follow. When he did so, puss led the way to the study-door, and going to a heap of books that lay on the floor, began to push her paws under them. Mr. Wood took up the books, one by one, the cat watching with looks of expectation; and when the last volume was taken up, forth darted a mouse, which the cat seized and despatched, and then began to purr, as if asking for congratulations and praise for its watchfulness.

Mrs. **Lee** tells us that her mother-in-law had two

great favourites—a cat and a canary. The two formed a friendship without the knowledge of their mistress, who was rejoiced to perceive their good understanding. Now they were allowed to be constant companions. But one day, while they were together in the bedroom, their mistress was alarmed by seeing puss, after uttering a menace, seize the canary in her mouth, and leap with it on to the bed, her hair bristling up as if in defiance. But the cause of this excitement was soon understood;—the door had been left open, and a strange cat had stolen in. It was to save the canary from this new peril that her friend had seized, in order to protect, him. When the stranger was driven away, the bird was set at liberty, and was found to be quite unharmed.

Mr. Jackson mentions a cat belonging to a widow at Stoke Newington, who always shut up her house and went away on Sundays. The cat did not like this solitary confinement; and she accordingly laid a plan, which she regularly carried out, for leaving the house every Saturday afternoon, paying a visit to the house of a neighbouring gentleman, and remaining there all Sunday, but returning home every Monday morning.

Mr. Wenzel had a cat who had formed a strong friendship for a dog, from whom she would never willingly be separated.   They were accustomed to share their meals, to sleep close by each other, and to go abroad together.

One day, Mr. Wenzel, to test the cat's friendship, took her into his dining-room, and gave her a share of the dinner, keeping the dog excluded.   He watched her, and found that she seemed to enjoy her dinner without a thought of her companion.   He then went out, leaving half a partridge, which was reserved for his supper,—his wife covering it up in a cupboard, the door of which, however, she did not lock.   But the cat, when left at liberty, went in search of the dog; and when she found him she uttered several expressive mews, which he answered by short barks. They then went together to the door of the dining-room, waited till some one opened it, and crept in both together.   The cat led the way to the cupboard where the half of the partridge was, and pulling open the door, pushed off the cover, and showed the feast to her friend, who eagerly seized and devoured it.

Mr. Jesse says that a man who had been sentenced to transportation for robbery told him, after his conviction, that he and two others broke into a gentleman's house near Hampton Court.   While they were engaged in carrying off the valuables, a large cat flew at one of them, and fixed her claws on each side of his face.   The

PUSS AND PUG.

man said that he never saw any one so much frightened as this burglar then was.

Often, when they do agree, the attachment of the dog and cat is something very remarkable.

Mr. Wood tells us that in a large London house there had been a change of servants, and the new cook asked as a favour that she might be permitted to bring in her favourite dog. Permission was given, and the dog took up his quarters in the kitchen—at first, to the infinite disgust of the cat, who disliked this introduction of a stranger and a rival. But after a time puss got over her dislike, and the two animals became fast friends. Another change took place : the cook removed into another family, and took her dog with her. One day, some months after, she determined on paying a visit to her former companions, and her dog went with her. Pussy was in the kitchen when the dog entered, and at once flew to greet him. She then ran out of the room, and shortly afterwards returned, bringing in her mouth her own dinner. This she laid down before her old friend, and actually stood beside him while he ate the food on which she had intended to make her own meal. This fact was related to Mr. Wood by the cat's mistress.

"An Irish gentleman," says Mr. Chambers, "removed his establishment from the county of Sligo to a house near Dublin,—a distance of not less than **ninety** miles. In making this change, he and his children regretted very much being obliged to leave a favourite cat behind them, which had endeared itself to them by its docility and affection. This gentleman **had** not been many **days in** his new abode, when, one evening, as the family were sitting chatting after tea, **the** servant came in, followed by a cat so precisely like the one left behind, that all the family cried out at once. The cat, too, showed the greatest joy **at** the meeting. It was closely examined, and **was** evidently the same. Still, how could it be **their own pet?** for how could he have found **them** out? how could he have made the journey? how **could he** have known where they were? Yet the exact-**ness** of his resemblance, and his evident joy at the meeting—walking about with tail erect, and loudly purring, made it impossible to doubt the identity. The gentleman took him on his lap, and, examining him closely, found that his **claws were actually** worn down,—making **it** evident that he had really performed that great journey."

" A kitten," says Mr. Jesse, " had been put into a pail of water in the stable-yard of an inn, for the purpose of drowning it. It had remained there for a minute or two, until it was to all appearance dead, when a female terrier, attached to the stables, took the kitten from the

water and carried it off in her mouth. She soon
brought it to, suckled and watched over it with great
care ; and it lived and thrived. She had at the time a
puppy of her own."

A lady writes to Mr. Wood, describing "a lovely
kitten, named 'Pret,'" of whom she says that "she was
the wisest, most loving, and dainty pussy that ever
crossed my path.

"When 'Pret' was very young I fell ill with a
nervous fever. She missed me immediately from my
accustomed place ; sought for me, and placed herself at
my door until she found an opportunity of getting into
the room, when she began at once to try her little best
to amuse me with her little frisky kitten tricks and pussy-
cat attentions. But soon finding that I was too ill to
play with her, she placed herself beside me, and at once
established herself as head-nurse. In this capacity few
human beings could have exceeded her in watchfulness,
or in affectionate regard. It was wonderful how soon
she learned to know the different hours at which I ought
to take medicine or nourishment; and if, during the
night, my attendant slept, she would call her, and if she
could not awake her without, would gently nibble her
nose, which always roused her. This done, 'Miss Pret'
would watch the preparation of the food or medicine,
and then come to me to announce its approach. The

strangest thing was, that she was never five minutes out
in her calculations of the true time, even in **the** darkness
and stillness of night.   There seemed, here, something
more than reason.

At the barn-yards of Castle Forbes several doors open
by means of the common thumb-latch, and these **a**
small grey cat had discovered the means of opening for
herself.   At various times pussy had been seen to
spring from the ground, a height of about four feet,
fasten her left fore-leg in the handle, and with her right
paw press the latch till she lifted the inside portion,
when the door swung round, and she dropped to the
floor at her own pleasure.   One of the doors is a heavy
panelled outside one.   On one occasion she was observed to fail in opening it; but, nothing daunted, she
made a second attempt, and crept up till she put her
whole weight upon it, and was rewarded with success.
It is supposed that her first **inducement** to attempt
the ingenious feat was for the purpose of visiting her
kittens, when the door happened to be shut.

The Rev. E. Spooner writes to the *Animal World*:—

" A few years ago we had a pretty kitten, for which
we were anxious to find a home.   A friend, who was
visiting us, gladly accepted the charge of her, and took
her **to** his own house.   There ' Miss Puss ' became **a**

great favourite; but her choicest affections were lavished
on one of the sons of the family—a remarkably bright
and handsome young man. It was interesting to see
the eagerness with which she watched for his return
from his office, the readiness with which she hastened
to meet him, and the happy composure with which she
took her place on his shoulder while he was at meals;
receiving morsels from his hands, rubbing herself against
his cheek, or playing with his whiskers. The friendship
grew so strong that puss would not leave her friend even
at bed-time, but took her place regularly on his pillow.
Some months had elapsed, when the young man was
suddenly missing. Many days of deepest anxiety were
spent before we discovered the sad truth that he had
come to a most melancholy and violent end. When the
poor body was found, it was impossible to take it home;
it had to be buried as quickly as might be: and never
can I forget that sad funeral. From the very day that
her young master was missing, poor puss seemed to
share the grief and anxiety of the family. She was
depressed and miserable, refused food, and wandered
about mewing piteously. At the time of the young
man's disappearance the family was on the very point
of moving to a new home; in fact, the move was begun
before the funeral and hurried on after it. When it was
nearly accomplished, pussy was taken by her mistress in
a cab through several unknown streets to her new domi-

cile. Here she seemed restless and **unhappy**. After
two or three days she was missing, and no trace of her
could be found. The family grieved for her; for her
affection for their lost favourite had attached them to her.
About a week after she had been lost, her master was
compelled to go to his old house to meet the landlord,
and to give up possession. Arriving there a little
before the time appointed, he thought he would just
look round the house and see if perchance aught had
been left behind. Wandering through the empty rooms,
he reached at last the upper floor; and there, on the
landing, just before his lost boy's bedroom door, lay the
dead kitten. I need add no comment."

Allowing the first place to the dog, man's constant
and favourite companion; and the second to the soft
and familiar cat, we must not forget those creatures of
a different kind, who find life tolerable in cages, and
who, thus enclosed, are found in so many of our dwell-
ings. Two of these, perhaps, will suffice,—the parrot,
and the song-bird.

Mr. Jesse speaks of a parrot which he had seen at
Hampton Court, and whose intelligence had so much
astonished him that he requested the sister of the lady
who owned it to furnish him with some particulars. She
wrote the following sketch :—

" As you wished me to write down whatever I could collect, I will now do so; only premising that I will tell you nothing but what I can vouch for, as having myself heard. Her laugh (the parrot's) is quite extraordinary, and it is impossible to help joining in it oneself; especially when in the midst of it she cries out, ' Don't make me laugh so! I shall die! I shall die!' and then continues laughing more violently than before. Her crying and sobbing, too, are curious; and if you say, ' Poor Poll! what is the matter?' she answers, ' So bad! so bad! got such a cold!' and then, after crying for some time, will gradually cease, and making a noise like drawing a long breath, will say, ' Better now,' and begin to laugh. The first time I ever heard her speak was one day when I was talking to the maid at the bottom of the stairs, and heard what I then took to be a child calling out, ' Payne (the maid's name), I'm not well! I'm not well!' and on my saying, ' What is the matter with that child?' she replied, ' It's only the parrot; she always does so when I leave her alone, to make me come back'; and so it proved, for on her going into the room, the parrot stopped, and began to laugh, quite in a jeering way. It is singular enough that, whenever she is affronted in any way, she begins to cry; and when pleased, to laugh. If any one happens to cough or sneeze, she says, ' What a bad cold!' One day, when the children were playing with her, the maid came into the room; and on their repeating to her several things which the parrot had said, Poll looked up, and said, quite plainly, ' No, I didn't.' Sometimes, when she is inclined to be mischievous, the maid threatens to beat her; and she says, ' No, you won't.' She calls the cat very plainly, saying, ' Puss! puss!' and then answers, ' Mew '; but the most amusing part is, that, whenever I want to make her call it, and for that purpose say, ' Puss! puss!' myself, she always answers, ' Mew,' till I begin mewing, and then she begins calling ' Puss' as quick as possible. She imitates every kind of noise, and barks so naturally that I have known her to set all the dogs on the parade at Hampton Court barking; and the consternation I have seen her cause in a party of cocks and hens, by her crowing and chuckling, has been the most ludicrous thing possible. She sings just like a child; and I have more than once thought it was a human being : it was ridiculous to hear her make a false note, and then cry out, ' Oh, la!' and burst out laughing at herself, beginning again in quite another key. She is very fond of singing ' Buy a broom?' which she says quite plainly; but in the same spirit as in calling the cat, if we say, with a view to make her repeat it, ' Buy a broom?' she always says, ' Buy a *brush?*' and then laughs, as a child might do when mischievous. She often performs a kind of exercise which I do not know how to describe, except by saying that it is like the lance exercise. She puts her claw behind her, first on one side and then on the other, then in front, and round over her head; and while doing so, keeps saying, 'Come on! come on!' and when finished, says, 'Bravo! beautiful!' and draws herself up. Before I was as well acquainted with her as I am now, she would stare in my face for some time, and then say, 'How d'ye do, ma'am?' This she invariably does to strangers. One day I went into the room where she was, and said, to try her, 'Poll, where is Payne gone?' and to my astonishment she replied, 'Down-stairs.' I cannot, this moment, recollect more, and I do not choose to trust to what I am told; but from what I have seen and heard, she has almost made me a believer in transmigration."

Mr. Smee tells us of a parrot belonging to a friend of his, which bird exhibited more than usual powers of thought and contrivance. He says,—

"I lent a book to a near relative, with whom this bird lived. On calling at the house, I found Polly sitting on the table; and I observed that she had torn the cover of my book to pieces. I felt angry; but, on ringing the bell, the servants assured me that the bird had been left shut up in its cage, and that she had opened the spring,—adding that they had lately observed that she had found some way of letting herself out.

6

We agreed that **this must** be stopped, **and it** was determined that a padlock must be added to the fastenings.
This padlock was opened by pressing a spring. The
next day Polly was again found outside the cage, **with**
the padlock lying at the bottom. She was soon put
back into the cage, and the padlock duly fastened; but
she walked deliberately down, took hold of the padlock,
opened it, and walked triumphantly out."

Mr. Bingley tells **us that,—**

"A pair of Guinea parrots were lodged together in
a large square cage. They usually sat on the highest
perch, and **close** to each other. When one descended
for food, the other always followed, and when their
hunger was satisfied they returned together to their
usual resting-place. They passed four years together
in this cage, and it was evident that a strong affection
existed between the two. But after **a** while the female
began **to** exhibit symptoms **of** old **age. She** could no
longer descend to take her food **as** formerly. The male
then assiduously attended **on** her, bringing food in his
bill and putting it into hers. He continued, in this
way, to feed her **with the utmost** care for about four
months. But her weakness increased, and at last she
seemed unable to keep her seat on the perch, but sank
down to the bottom of the cage, where she remained,
crouched up in **one of** the corners. She sometimes

seemed to wish to rise again, at least as far as to the
lower perch; and her partner eagerly seconded her
efforts, sometimes seizing her wing, sometimes her
bill, to try to draw her up to him;—his gestures and
eager solicitude showing his earnest desire to assist her
weakness and alleviate her sufferings. At length her
death evidently approached; and now he paced round
and round her without ceasing, redoubling his assi-
duities. He tried to open her bill, so as to give her
food; but all in vain; and now he paced to and fro
in the highest agitation. Sometimes he uttered plaintive cries; at other times he would sit for hours with
his eyes fixed upon her. At length she expired. He
languished for a few months, and then followed her."

We pass on to the finches and other song-birds, which
often impart a loveliness to a solitary chamber.

Mrs. Webber, in her work on " The Song-birds of
America," tells the following story, which we give
because she is narrating things within her own personal knowledge. She had lost a pet thrush, and
thought she should never love any other bird so much.
But a piping bullfinch was brought to her, and his
winning ways broke down her resolution :—

"Although I still said I did not love him, I talked a
good deal to the bird; and as the little fellow grew more
and more cheerful, and sang louder and oftener every

day, and was getting very handsome, I began to increase my attentions to him. He seemed, too, to need
my presence quite as much as sunshine; for if I went
away, he would utter the most piteous cries till I came
back, and then, in an instant, his tones were changed,
and he sang his most enchanting airs. He was most
devoted in his efforts to enchain me by his melodies,
and by degrees I began to express my admiration. He
then seemed satisfied ; but began to claim me wholly.
No one else must approach him, and if any one laid a
finger on me his fury was unbounded. Now, if I went
away he would first mourn, then attempt to win me back
by sweet songs. If I sat too quietly at my drawing, he
would become weary, and would call in gentle tones,
' Come-e-here! come-e-here!' so distinctly that all
my friends recognised the accents. All the day long he
would watch me: if I were cheerful, he would sing and
be gay; if I were sad, he would sit by the hour watching
every movement; and if I arose from my seat, I was
called, ' Come-e-here!' If I let him fly about the
room, he would follow me so closely that I was in constant terror lest he should be trodden upon. He now
wanted to feed me, like a young bird. But I did not
like this sort of relationship, and determined to break
it off. I refused to be fed, I turned away, and as fast as
I could, broke down all the gentle bonds between us.
The result was sad enough. The poor fellow could not

bear it; he sat in wondering grief. He would not eat.
At night I took him in my hand, and held him to my
cheek : he nestled closely, and seemed more happy,
though his little heart was too full to let him speak.
In the morning I scarcely answered his tender call,
'Come-e-here!' but I sat down to my drawing,
asking myself if I could be so cold much longer to
such a gentle and uncomplaining creature. I presently
arose, and went to the cage. Oh, my poor, poor bird !
he lay struggling on the floor. I took him out; I tried
in every way to call him back to life, but it was useless.
I saw that he was dying ; his little frame was even then
growing cold. I uttered the call he knew so well : he
threw back his head, with its yet undimmed eye, and
tried to answer; but the effort was made with his last
breath. His eye glazed ; his little heart was broken !
I can never forgive myself! To kill so gentle and so
pure a love as that ! "

In the *Zoologist* of 1848 Mr Duff writes :—

" I had a canary, between the wires of whose cage I
used to fix a piece of lump-sugar. One day it dropped
out, and when picked up, it was found to be quite wet
on one side. This surprised me, and I replaced it in
the cage with the dry side inwards, and determined to
watch the bird's proceedings. To my surprise, after a
few ineffectual attempts on the hard sugar, the bird went

to the water-glass, filled its bill, dropped the water on the sugar, and repeated this several times.   When the sugar was thus softened, it began to eat it.   Evidently

there was thought—there was a calculation of means, and a use of them."

II.

IN THE STABLE.

WE pass from out of the house to its nearest adjunct. In the neighbouring building we have creatures of a very noble kind; though, from the distance at which they are kept, we scarcely understand or appreciate the qualities they possess.

Assuredly, **much less than** justice is commonly awarded to the horse. **He is** not only one of the most beautiful of God's creatures, but he possesses peculiar excellences both of mind and heart. He has a **most** wonderful memory; indeed, in this particular we have no other creature that can equal him. He has also intelligence of a very high degree; but, above all, he manifests, continually, both sympathy and attachment. Yet, because we **find it** convenient to place him in an out-house of his own, and to hold no converse with him, he comes, very soon, to be regarded as a mere "beast of burden," for whom we have scarcely more care than we have for the plough or the waggon which we compel him to draw.

This is a wrong, an injustice, which **ought to** be, and which can be, forsaken. Our present work, however, is **not** meant to be an *argument;* and we shall content ourselves, in this **as** in other matters, with the collection and arrangement of facts. The conclusions to be drawn from those facts we shall leave **to** the consideration **of our** readers.

The horse has a most retentive **memory,** and is susceptible of strong attachments. **Proofs of** this are abundant on **every** side.

"One day," **Mr. Morris writes, "a** gentleman was travelling **by a** stage-coach, and when he **got** down **at the inn, the coachman saw** that one of the leaders was trembling all over, making a little neighing noise, **and** looking round after the passengers. 'I think that horse knows you, sir,' said the coachman. The gentleman looked more attentively, and soon saw in the horse an old servant, whom he **had parted with** a few years before, **and** who **now showed these** marks **of pleasure on** meeting him again.

"I once had a handsome black mare, called 'Bess.' Whatever road I might intend to travel, she always seemed to be aware of it beforehand. And if I sometimes stole quietly into the stable, after a day's ride, to see if my man was looking properly after her, he would tell me how useless was the attempt. 'Why, sir,' he said, 'it would make no difference if you came without

your shoes: I should know long before you came near
the stable; for when you are coming the mare always
begins kicking and prancing like mad.' "

A curious case was brought before the magistrates
at Worship Street recently. A chesnut horse was stolen
from Hackney Marshes, and the thief (or some one to
whom he had sold it) happened to lead the horse past
his master's house in Old Ford, Bow. As he passed the
house, the animal neighed; and the owner, with whom
the horse was a favourite, knew the neigh, looked out of
window, and saw his horse being led down the street.
He instantly called for help, and the horse was rescued.

Mr. Smiles, in his "Lives of Engineers," tells us that—
"The whole of the stone for Waterloo Bridge (except the balustrades) was hewn in some fields adjacent,
on the Surrey side. It was then transported, stone by
stone, upon trucks drawn along railways, over temporary bridges of wood; and nearly the whole was
thus transported by one horse, 'Old Jack,' a most
sensible animal and a great favourite. His driver was,
on the whole, a steady man, though rather too fond of
his dram before breakfast. The railway along which the
truck passed, ran just in front of the door of a public-
house; and here Tom, the driver, usually pulled up for
his ' morning.' On one occasion it happened that Tom

stopped longer than his usual time, and 'Old Jack' grew impatient. So he pushed his head against the public-house door, got it inside, and finding his master standing at the bar, took the collar of his jacket between his teeth, and pulled him out of the place, back to his work again."

Mr. Banvard writes,—

"Returning from a tour to the West, I put up at a small town near the Alleghany Mountains. While I was sitting watching the variegated hues produced by the rays of the setting sun upon that wild, rough, mountain scenery, I saw eight or nine large baggage-waggons approaching, drawn by four, and some of them by six horses. I found that the tavern where I was stopping was a regular lodging-place for those strong, coarse, mountain-waggoners. Near the place where I sat, in front of the house, was a pump with a large trough, which was used for watering horses. The handle of the pump, I observed, always sprang up whenever any one used it. Most pump-handles fall down; but this one sprang up, so that those who used it had no occasion to lift the handle; it always raised itself. When the string of waggons approached the tavern there was but little water in the trough,—not nearly enough to supply the horses. But you may imagine my surprise to see one of the horses, as soon as he was unharnessed, go to the

THE THREE FRIENDS.

pump, lay his head over the handle, and press it down, so as to make the water come out of the spout. Then he raised his head, and the handle sprang up; then again he would press it down, and send more water into the trough. In this way that horse continued pumping, till all the horses had had the water they wanted. Then, last of all, he left the handle, walked round to the trough, drank as much as he wanted, and finished by walking into the stable and taking his place in one of the stalls."

A farmer who lives in the neighbourhood of Bedford, and regularly attends the markets there, was returning home one evening in 1828, and having drank till he became sleepy, rolled off his saddle into the middle of the road. His horse stopped, and waited some time, until, seeing no disposition in its rider to remount, it took him by the collar and shook him. The farmer only grumbled, and remained in slumber. The horse then began to resort to stronger measures—seizing hold of his master's coat-laps, and dragging at them to make him get up. After a while the coat-lap gave way. While this was going on, three more passengers came by, and these succeeded in awakening the farmer and inducing him to remount. The torn coat-lap was carefully thrust into the pocket, and was long preserved by the farmer, as a memorial of his horse's zeal and care.

A gentleman of Bristol had a greyhound, who slept in the stable **along** with a very fine **hunter.** Between the two a strong friendship was formed. The **grey-** hound always slept within sight of the horse, who **would** show signs of restlessness whenever his friend **was** absent. The owner would often call at the stable **and** summon the greyhound for a walk ; and the horse would look round and **neigh, in a** manner which plainly said, "Let me go with you." And when the dog returned home, he was always welcomed by a loud neigh ; when he would **run up to the** horse and **lick his nose ; and** the horse, in **return, would** scratch the **dog's back with his** teeth. One day, when the groom was out with them both for exercise, a large dog attacked the greyhound, and, by mere strength, threw him on the ground. The horse at once rushed to his assistance, seized the assailant by his back, and tore out a large piece of skin, compelling him **to a speedy retreat.**

The **island of Krutsand, which is** formed by two branches of the Elbe, is peculiarly liable to floods, caused by spring-tides, when the wind is in an un- favourable quarter. On one occasion the water rose so rapidly, that a number of horses that were grazing in the meadows, with their foals by their sides, found themselves, in a very short space of time, surrounded on all sides with deep water, so as to endanger the young animals. They soon all assembled together, and in this danger, anxious for their foals, they devised the only available plan. Each two horses took a foal between them, and pressing close together, lifted it up, and kept it partly above the water till the ebb came, and the danger gradually passed away.

Dr. Smith, of the Queen's County, Ireland, had a beautiful hackney, which, though extremely spirited, was at the same time wonderfully docile. He had also a fine Newfoundland dog, named "Cæsar." The two animals always slept in the same stable, and soon became close friends. When the doctor took his rounds, he needed no other servant than "Cæsar," leaving the horse in his care when he entered a patient's house. Sometimes he would walk a short distance, from one house to another, and tell the two friends to follow him. "Cæsar" would keep the reins in his mouth, and the two would walk quietly together after their master. Sometimes the doctor would go to the stable, put the reins upon his horse, and then giving them to "Cæsar," would bid him take the water. They both understood the order, and went together to the rivulet, about three hundred yards off. The horse quenched his thirst, and then he and "Cæsar" returned to the stable. Sometimes the doctor would bid "Cæsar" make the horse leap the stream, which was not very wide. The dog knew

how to make the horse understand what he was to do;
and they both cantered off, and the horse took the leap
in very good style. When "Cæsar" lost the reins,
which would sometimes happen, the horse would trot
up to his friend, and allow him to take them again in
his mouth.

In the days of George III., when volunteer corps were
embodied in every considerable town in England, a new
line of turnpike-road was in course of construction
through a northern county. The duty of one of the
clerks to the trustees was to superintend the contractors and their men, and to see that the work was done
properly. When taking these rides along the line, he
generally rode an old horse which had formerly carried
a field-officer, and still possessed much courage and
spirit. One day, while taking one of his journeys of
inspection, the clerk passed near a town of some impor-
tance, where, at that moment, the volunteers were out
for drill on the common. No sooner did the horse hear
the drum, and the word of command, than he cleared
the fence, and was soon in the commanding officer's
place, in front of the volunteers; nor could his rider
induce him to move from that position till the volunteers had ended their evolutions, and begun to march
off. He then marched at their head into the town,
prancing proudly in martial recollection of former

days; to the great amusement of the bystanders, but somewhat to his rider's annoyance,—who was thus brought into an absurd position against his will.

A cart-horse belonging to Mr. Leggatt, of Glasgow, was troubled with a disease, for which he had been treated by a farrier of the name of Downie. For a considerable time he had not been troubled with his old complaint; but one day he was employed in College Street, Glasgow, at a distance of nearly a mile from the farrier's shop; and while mixed with the other horses, the carters being busy elsewhere, he was missing. People were sent in all directions in search of him, but in vain. He had quietly taken his departure for the farrier's shop—where he presented himself, to the surprise of the people, without any attendant. They surmised, what proved to be the fact, that he felt his old complaint, and was in search of a cure. They took off his harness, and he then laid down and gave signs that he was suffering. The usual remedies were administered, and he was sent home to his master, who was in the greatest perplexity as to what had become of him.

"A friend of mine," says Mr. Morris, "was riding home one night through a wood; and, owing to the darkness, struck his head against the branch of a tree, and fell from his horse quite stunned. The horse immediately returned to the house which he had just left—a mile or more distant. He found the doors closed, and the family retiring to rest. He pawed at the door, until one of the servants, hearing a noise, came down and opened it ; and saw, with surprise, the horse which had so recently left them. But the creature, so soon as he was recognised, turned round, inviting the inmates to follow. They did so, and the creature led them straight to the spot where his rider still lay upon the ground, stunned, and scarcely able to rise."

In Fifeshire, a carter at Strathmiglo had an old horse, who had long lived with him, and was intimate with all the family. When among the children, he was always careful to do them no injury—moving his feet with the greatest care. On one occasion, he was drawing a loaded cart through a narrow lane, when he found a young child playing there, and liable to be crushed by the cart-wheels. He took the child up by the clothes with his teeth, and carried it a little way, till he saw a bank by the road-side, on which he could safely place it. Then, having thus put it out of danger, he went on his way; but not without looking back, to satisfy himself that the wheels of the cart had cleared it.

The Shetland pony is a creature more full of fun and frolic than any other.

"Ours," says Mr. Morris, "could open every gate which hindered his rambles. There was scarcely any common fastening which he could not undo with his teeth; and if foiled here, he would find out the weakest place in the fence, and break through it. His only object was, greater freedom for an hour or two's gallop. With equal adroitness, he would find out and open the grain-chest in any stable. When occupied with any mischief of this sort, he would be shy and distant; while at other times he would follow like a lapdog, asking for apples or bread. When idle in the paddock, nothing pleased him more than a game of romps with any rollicking dog. Once a day the mail-coach came by, and when he heard the horn he always ran to a place where he could see it. As soon as it appeared, he would run round and round his paddock, his heels usually higher than his head, and his mane and tail streaming out, as if he were showing himself,—to the amusement and admiration of the passengers."

Some horses can be managed by kindness, but in no other way. A horse in the depôt at Woolwich was so unmanageable that, at length, no one, even of the rough-riders, durst venture to mount him. He would either crush his rider's leg against some post or wall, or would lie down and roll over him. No means could be found, of curing him of these tricks; and at last he was reported to the commanding-officer as " incurably vicious," in order that he might be condemned to be sold, and turned out of the service. But Col. Quest, admiring the horse, which was thorough-bred, and of fine action, asked the commanding-officer to hand him over to the riding-troop. This was done; and Col. Quest resolved to try a different system of management from that which was usually adopted in the riding-school. He had him daily paraded, took notice of him, patted him, and kept all whips out of sight. When the horse did anything as ordered, the colonel gave him a piece of bread, or a handful of beans. By this sort of treatment he soon gained the creature's confidence and good-will; and the horse, at last, became so docile that a child might be placed on his back. He would even kneel down to be mounted, and would perform evolutions which no other horse seemed able to understand. At last, such a favourite did he become, that his master gave him the appellation of "Darling."

During the war in Spain a regiment of cavalry was ordered to embark from Plymouth Dock for the Peninsula. Amongst the horses was an old campaigner, which had been, it was said, more than once on the same errand, and appeared to have made up his mind *not to go on foreign service.* In pursuance of this deter-

mination, he resisted with all his might every attempt
to sling him on board the ship, kicking and plunging so
furiously that the men employed at length gave up the
attempt in despair. A resolute fellow of a sailor, seeing
how the matter stood, came forward, vowing he would
conquer him, and instantly grasped the horse round the
neck, with the design of fixing the necessary apparatus.
Jack, however, reckoned without his host : the horse, by
a sudden plunge, shook him off, and, turning his heels,
gave him a severe kick, which laid him sprawling on
the ground; and galloped off. But, after making a
circle, he returned to the spot where his antagonist lay,
and fairly hurled or pushed him into the water—out of
which he was dragged by the crew of a boat which was
near the spot.

A gentleman, on one occasion, rode a young horse,
which he had bred, to a distance of thirty miles from
home, and to a part of the country where he had never
before been. The road was a cross one, and extremely
difficult to find ; but, by dint of perseverance and inquiry,
he at length reached his destination. Two years afterwards he had occasion to pursue the same route. He
was perplexed three or four miles from the end of his
journey. The day grew so dark that sometimes he could
scarcely guess his way. He had a black and dreary
moor to pass, had lost all traces of his route, and even

felt uncertain whether he was going in the right direction. The rain, too, began to fall heavily. He be-
thought him of what he had heard of the wonderful
memory of the horse. He determined to throw himself

upon this, as his last resource. He threw the reins on
his horse's neck, and by word and action encouraged
him to go forward. It was not long before he found
himself at his friend's door. Yet it was absolutely certain that the horse could not have traversed that road a

second time since his first visit to that house, at a distance of two years previous.

Sir Emerson **Tennant,** writing to *Land and Water,* says,—

"A **few** days ago my attention was attracted to a **pretty** little horse, with a milk-cart, trotting across the **road,** and drawing up at a particular door. Presently the cart was followed by the man in charge, **who** had no sooner put himself in communication with the person inside than the horse went forward and drew up **at a** house a little further on, which was soon opened to him. I asked the driver whether the **horse** knew every house in his walk. He **answered, 'Yes, he** knew them all. If **a** new customer **was added, the horse soon** included him ; **and** if one **left, the horse** quickly **found it** out, and passed the door unnoticed.' I sought out **the** owner of the cart, and asked him concerning this interesting little animal. He told me that he had had him for **about five** years ; that he bought him cheap, because he bore the **character** of **being** vicious, and a jibber. But he tried the experiment **of** kindness, treating the horse sometimes to a slice of bread, and sometimes to a lump of sugar. The creature soon showed his pleasure and his gratitude, but seemed most pleased by a chesnut when **one** was given him. And now he showed no objection **to** the cart ; but rather a wish to contribute

his share in the service. The idea of moving in advance
of the driver was entirely his own. All that the driver
had to say was, ' Go on, Jerry!' and away the horse
trotted to the next place. Sometimes, too, he would do
more. I saw him myself (says Sir I. E. Tennant) trot
up alone to a door in Coburg Row, Pimlico. I saw him
cautiously approach the kerb-stone, taking care that the
wheel of the cart did not get upon the footpath; and
then with his nose he raised the knocker, and let it fall,
and then a second time; and then he waited till the door
was opened, and he was rewarded by a slice of bread,
and a pat on his neck while he was eating it."

Mr. Lane, of Frescombe, in Ashelworth, Gloucestershire, had sent his horse to be shod, and had then turned
it into a field to graze. The following morning it was
missing, and a hue and cry was set up. After a while
the state of the case was discovered. The farrier had
pinched the horse's foot in the shoeing. The animal,
feeling itself in an uncomfortable plight, escaped from
the field by lifting the gate off its hinges. It then went
straight to the farrier's, a distance of a mile and a half.
When it arrived there, it held up the ailing foot, and
showed what was its cause of complaint. The farrier
soon perceived how the matter stood, took off the shoe,
and replaced it more carefully. The horse tried it, and
was satisfied, and turned round and set off home. An

hour after, Mr. Lane's men, who had been scouring the country, called at the farrier's shop, and were told at once, "Oh, he has been here, has been re-shod, and is gone home again."

"Most domestic animals," says Mr. Hamerton, "are as keenly alive to their own interests as a man of business. They can make bargains, and stick to them, and make you stick to them also. I have a little mare who used to require six men to catch her on the pasture ; but I carried corn to her for a long time without trying to take her, leaving the corn on the ground. Next, I induced her to eat the corn while I held it, still leaving her free. Finally, I persuaded her to follow me, and now she will come trotting half a mile at my whistle, leaping ditches, fording brooks, in darkness, or rain, or fog. She follows me like a dog to the stable, and I give her the corn there. But it is a bargain : she knowingly sells her liberty for the corn. The experiment of reducing the reward having been tried, she ceased to obey the whistle, and resumed her former habits. But the full and due quantity having been restored, she yielded her liberty again without resistance ; and since then she is not to be cheated."

"A farmer's boy," says Mr. Jesse, "had fed and taken great care of a colt. One day, while working in the field, he was attacked by a bull. The boy leaped into a ditch, to get out of the animal's way. But the bull would not leave him, and tried to reach him with his horns, and would probably have succeeded, had not the colt come to his assistance. The colt not only rushed at the bull, and kicked him, but made so loud a noise that some people who were within hearing came running up to see what was the matter, and contrived to drive the bull away."

In another case, a horse and a cat became great friends, insomuch that the cat generally slept in the horse's manger. But when the horse received his corn, and was about to take a meal, he always took up the cat by the skin of her neck, and dropped her into the next stall, that she might not be in his way, and get into any danger while he was feeding.

"Two Hanoverian horses," says Mr. Jesse, "had long served together in the Peninsular war, in the artillery of the German Legion. They had always drawn the same gun, and been companions in many battles. One of them, at last, was killed in action, and the survivor was left alone, picketed as usual, but without any companion. His food was brought to him as usual, but he refused to eat. He continued restlessly to turn his head in every direction, often neighing, as if to call his comrade. Attention was directed to him, and great care was taken of him. Other horses surrounded him on every side, but he took no notice of them. His affliction was apparent to all. He persisted in refusing to touch food, and after lingering a few days, he died, to the grief of all those who had watched him."

A writer in *Science Gossip* says,—

" Last year, being on a tour in Bedfordshire, I stayed at a friend's house, where a horse, in the back-yard, came to the door of the room where I was sitting, took in his mouth the handle of it, and by a twist of his head opened the door and put his head into the room. The mistress of the house understood what he wanted, and put a lump of sugar into his mouth; when he at once backed out of the room, closing the door after him."

A writer in the *Naturalist's Magazine* says,—

" An old mare belonging to a man in my village, which looked as if it had hardly sense to do its work, had a foal last summer; and, one day, the mother came galloping up the village to its owner's door, neighing and showing great agitation. The man said, ' Something must be the matter,' and he went out to her. She at once trotted off, looking to see that he followed her; and she led him to the mill-dam, where he found her foal, who **had** slipped in, and was in danger of being drowned.''

" We knew a horse," says the author of " The Menageries," "who, being accustomed to be employed once **a week** on a journey with the traveller of a provincial paper, always stopped at the **houses** of several customers, sixty or seventy **in** number.   **Now, there** were two of the customers who took the paper between them—the one claiming to have it first *this* week, the other the following.   The horse soon came to understand this **arrangement;** and though the parties lived two miles apart, he regularly stopped, one week at Thorpe, the other week at Chertsey—in no one instance mistaking the week, though the arrangement lasted for years.''

"A horse," says Mrs. Lee, " **had** been in **the habit** of going with his master a certain **road, and** stopping at a certain **inn,** where the hostler, **perhaps** by order, alw**ays threw** some beans **into the** corn supplied to him. After a time, the horse, with his master, removed from that part of the country, and remained away for two years.   After this, they returned, and came back to the same inn to which the horse had been accustomed. This inn, however, had passed into new hands.   While enjoying his dinner, the rider was informed that his **horse** would **not** eat.   He seemed **to** think that there

WAITING FOR THE MASTER.

was something wrong about the corn; but they knew that it was of the very best. The rider went to the stable; the horse neighed, and looked at the manger, and then at him. He recollected. 'Throw some beans in,' he said. The hostler obeyed; the horse looked at him as if to express his thanks, and then took his meal contentedly."

Mr. Jesse gives the following incident, as coming within his own knowledge :—

" A farmer residing on the borders of the New Forest, in Hampshire, went over to the Isle of Wight, where he purchased a mare from a person with whom he was acquainted, near Newport, in the centre of the island. The mare was conveyed over the water in a boat, landed in Hampshire, taken to the purchaser's farm, and turned into one of his fields. The next morning she was nowhere to be found. Search was made in all directions, but nothing could be heard of her, and it was supposed that she must have been stolen. But the farmer, having again occasion to go to the Isle of Wight, called upon his acquaintance from whom he had bought the mare, and learned, to his surprise, that she had returned safely to her old quarters. Over and above the land journey, which was considerable, she must have swum at least five miles, or more, to reach the island from the Hampshire coast."

Samuel **Drew**, in his "Autobiography," **says,—**

"My father had contracted to carry the **mail on** horseback between St. Austell and Bodmin, which duty commonly devolved on my elder brother Jabez. At **one** time, in the depth of winter, Jabez being ill, I (then ten years old, apprenticed to a shoemaker) was borrowed to supply his place, and had to travel in the darkness **of** night through frost and snow, a dreary journey out and home of more than twenty miles. Being overpowered with fatigue, I fell asleep **on** the horse's neck, and when I awoke observed that **I had** lost my hat. **The** wind **was keen and piercing, and** I was bitterly cold. I **stopped the horse and** endeavoured to find out where I was; but it was so dark that I could scarcely distinguish the hedges on each side of the road, and I had no means of ascertaining how long I had been asleep or how far I had travelled. I then dismounted, and looked round for my hat, but seeing nothing of it, I turned back, leading the horse, determined to find it, if possible. The loss of a hat to me was a matter of serious consequence; since, if it were not recovered, I should probably have to wait a long time for another. Shivering with cold, I pursued my solitary way, scrutinizing the road at every step, until I had walked about two miles; and was on the point of giving up the search, when I came to a receiving-house, where I ought to have delivered a packet of **letters,** but had passed it when asleep. To this place the post usually came about two o'clock in the morning, and it was customary to leave a window unfastened, except by a large stone outside, that the family might not be disturbed at so unseasonable an hour. I put my letter-bag through the window, and having replaced the stone, was turning round on my horse, when I perceived the hat lying close to my feet. I suppose that the horse, knowing the place, must have stopped at the window for me to deliver my charge, and that my hat was shaken off by his movements or efforts to waken me. Not succeeding in this, and having waited until his patience was exhausted, he had, though blind, pursued his way to the next stopping-place. By all the family this sagacious and valuable animal was much prized; but my father felt for it an especial regard, and the attachment between the master and his faithful horse was to all appearance mutual. Many years before, the poor beast, in a wretched condition from starvation and ill-usage, was turned out on a common to die. The owner willingly sold it for little more than the value of the skin, and his new possessor having, by care and kindness, restored it to strength, soon found that he had made an advantageous bargain. For more than twenty years he and his blind companion travelled the road together, and many were the proofs of its intelligence and attachment. After the horse was past labour it was kept in the orchard, and attended with almost parental care.

Latterly it had become unable to bite the grass, and the old man regularly fed it with bread soaked in milk. When in the early morning the horse put his head over the orchard railing, towards his master's bedroom, and gave its usual neigh, he would jump up out of bed, open the window, and call to it, saying, ' My poor old fellow, I will be with thee soon.' And when the animal died he would not allow the skin or shoes to be taken off, but had the carcase buried entire."

A correspondent of the *Animal World* says, —

" A young gentleman named Keene has been for some time resident near Malton; and from Mr. Rutter, of Hessle Farm, he bought a hunting-mare, which, on leaving Malton, he recently took with him to Whitby. On Wednesday the mare was missing from the field, and a search was instituted, to no purpose. On Thursday the search was renewed, Mr. Keene and his groom going about ten miles on the Guisborough Moors, and then to Sleights, where they heard the mare had crossed the railway the previous morning. At this point the trail was easy. The mare had taken the high road homewards, and at Saltersgate six men tried to stop her, without avail. At Pickering she jumped a load of sticks and the railway-gates, and then found herself in her old hunting country, making across Ryedale for ' home.' In so doing she would have to cross two rivers and a railway. Mr. Keene found her at home on Thursday night with one **shoe** thrown, and rather lame, but otherwise no worse for her cross-country gallop of nearly sixty miles, done in one day; for her previous owner found her on the Wednesday night standing at the gate of the field where she had grazed for the two previous years."

Another correspondent of the same journal says,—

"Directly opposite my residence a church is being erected, and during its progress temporary sheds have been put up for the use of the workmen, and one as a stable for a very fine cart-horse, the property of the builder. The extreme docility of this animal attracted my attention to him, **and** since **that some of his** manœuvres appear **to me to** border **strongly on the sense** and the powers **of** reflection. His stable **was erected at** one end of the church. On one occasion **two poles had** been fastened across his usual road to it, in order to strengthen the scaffolding; he went up, tried the strength **of these first, then** finding that he could neither get over nor under, he turned round, and, at a full trot, made **the** circuit of the church, and got to the other side **of** the poles by another path. Here was no straying about, **and at** last finding his way; he resolved to go round, as **if an** idea had at once flashed across his mind. Another **day, a waggon** had been left standing **in the** narrowest part of his road to the stable: he looked, and tried each side, but found **there** was not space enough **for** him to pass; he took very little time for consideration, but put his breast against the back-part of the waggon, and shoved it on to a wider part of the road, then deliberately passed on one side to his stable. Could human wisdom have done better? But, to crown all his manœuvres, I mention the following as being, I consider, very extraordinary:

"During the winter a large wide drain had been made, and over this strong planks had been placed for our friend, the cart-horse, to pass over to his stable. **It had snowed during** the night, and froze very hard in **the** morning. How he passed over the planks on going out to work I know not, but on being turned loose from the cart at breakfast, he came up to them, and I saw his fore-feet slip; he drew back immediately, and seemed for a moment at a loss how to get on. Close to these planks a cart-load of sand had been placed; he put his fore-feet on this, and looked wistfully to the other side of the drain. The boy who attends this horse, and who had gone round by another path, seeing him stand there, called him. The horse immediately turned round, and set about scraping the sand most vigorously, first with one foot, then the other. The boy, perhaps wondering what he would be at, waited to see. When the planks were completely covered with sand, the horse turned round again, and unhesitatingly walked over, and trotted up to his stable and driver."

Mr. W. S. Bellows, in the *Naturalist's Magazine*, writes,—

"When a boy, being at Whitchurch, near Blandford, I noticed two cart-horses that were sent from a farmyard to drink at a neighbouring brook. The brook was frozen over, so one of the horses struck the ice with his foot, to break it; but it was hard, and did not yield. The two horses, then, standing side by side, lifted each a foot, simultaneously, and making their hoofs descend together, the double blow broke the ice. Men could not have done better.

The *Scotsman* newspaper, on the authority of several trustworthy witnesses, mentions a circumstance which had occurred near West Calder. During a great heat which prevailed one summer's afternoon, a pony, the property of Mr. John Waddell, contractor, was left alone for some time by its driver. Having been driven a considerable distance, and feeling, naturally, a craving for water, the pony was observed by several persons to walk, deliberately, about fifty yards, and with its teeth turn the cock of a water-pipe projecting out of the embankment near, and having thus quenched its thirst, to turn the cock back again, so as to prevent

8

any waste, and then to go back to the place in which
its owner had left it.

Mr. Morris tells us of a Norwegian pony which
showed more than usual skill and adroitness. She was
brought over, with a companion, and at first they were
" turned out to grass," and enjoyed an entire holiday.
To this they soon added entire liberty, for they could
unfasten, undo, or untie every gate, and wander where-
ever their inclinations led them.   They soon had a companion, a foal of last year, which, being of great beauty,
was made much of, and had two feeds of corn every
day. The pony, however (the one particularly spoken of)
did not understand this partiality, nor would she submit
to it. She always contrived to get into the foal's shed
at his dinner-time, and to share the corn with him. If
the groom tied the door with a stout rope, she could
untie the knot with her teeth; if he fastened it with a
chain and staple and wooden peg, no sooner was his
back turned than the peg was drawn, and the door
opened. At length, tired with being so often beaten,
he got a heavy rail, which he placed right across the
entrance. The pony was puzzled, and the groom
looked on with glee, while he saw her vain attempts to
lift the rail, which proved too heavy for her. He
thought that the victory was his, when she seemed to
give it up in despair, and trotted off to find her com-

panion. But not a little astonished was he to see her
return in a few minutes, and the other pony with her.
Together they put their necks under the rail; and what
was too heavy for one, yielded to the efforts of both : the
rail was lifted and thrown down, and the way to the
corn was again opened. On another occasion, when
shut up in a yard, which they did not at all approve,
all kinds of fastenings gave way to them, till at last
the groom, in despair, actually nailed up the gate with
some stout tenpenny nails !

It would be unjust to pass over in silence the horse's
near relatives—the donkey and the mule. One of the
greatest, and yet the most common, of all mistakes we
make in every-day life is to deem the ass a stupid
animal. Let the following anecdotes form the best
reply to this groundless supposition.

The Rev. C. Otway writes,—

" I assert that if you were to make yourself acquainted
with asses, you would find them clever enough. I once
bought an ass for the amusement of my children. I did
not allow him to be cudgelled, and he got something
better to graze upon than thistles. I found him more
knave than fool; his very cleverness was my plague.
My ass, like the king's fool, proved the ablest animal

about the place ; and, like others, having more wit than
good manners, he was for ever not only going but
leading other beasts into mischief. There was not a
gate about the place but he could open it ; there was
not a fence that he could not climb. Often he would
wake me on a summer's morning, braying, out of sheer

wantonness, in the middle of a field of wheat. I was
obliged to part with him and get a pony, merely
because he was too cunning to be kept."

In March 1816, an ass belonging to Captain Dundas,
R.N., then at Malta, was shipped on board the *Ister*
frigate, Captain Forest, bound from Gibraltar for that
island. The frigate struck upon some sands off the
Point de Gat, at some distance from the shore ; and the
ass was thrown overboard, to give it a chance of swimming

to land—a poor chance, for the sea was running so high
that a boat which left the ship was lost. A few days
afterwards, however, when the gates of Gibraltar were
opened in the morning, the ass presented himself for
admittance, and proceeded at once to a stable belonging to Mr. Weeks, which he had formerly occupied.
Mr. Weeks was astonished to see him, and supposed
that, by some accident, the animal had never been
shipped on board the *Ister.* But when the vessel
returned for repairs, the mystery was explained, and it
was made manifest that the ass had not only swum safely
to shore, but had found his way from Point de Gat to
Gibraltar, a distance of more than two hundred miles,
through a mountainous and difficult country, intersected
by streams; and this in so short a period that it was
clear he could never have made one false turn.

Mr. East writes,—

"I had once a donkey, presented to me by Osgood
Hanbury, Esq., which was a remarkably docile and
knowing animal. He was the constant companion of
my children in their rambles on the Downs, and always
seemed to think that he had a right to his share in all
the eatables and drinkables, whether cakes, apples,
oranges, milk, beer, or even tea. Ginger-beer was
the only thing he eschewed. A ginger-beer cork had
once struck him on the nose, after which he would turn

his back when such a thing was produced. His lodging-place at night was in a shed, from which he had free access to a yard ; but not, of course, to the kitchen-garden, which adjoined it. This garden was **separated** from the yard by a wall, in which was a door, or gate, fastened securely by two bolts and a latch. But soon **we** were surprised to find that the door had been opened in the night, and there were footprints of the donkey on the walks and beds. How this could be we could not imagine, especially as the upper bolt was fixed at a considerable height. So I watched at my window, **and** saw master donkey, reared up on his hind-legs, unfastening the **upper** bolt with his mouth. He then drew back the lower one, lifted the latch, and walked quietly into the garden. In a few minutes he returned, bringing a large bunch of carrots, which he deposited in **his** shed, and then went back to latch the gate, after which he leisurely set about munching up his booty. Before putting a stop to these proceedings, I gave some of my neighbours, who were incredulous on the subject, an opportunity of witnessing them. It should be added that master donkey never commenced his operations until after the light had been extinguished at the bedroom window."

The *Kelso Mail* gives the following brief narrative :—

A ROUGH LOT.

" T. Brown, residing near Hawick, travels the country as higgler, or pedlar, having an ass for the partner
of his journeys. Weakened by a touch of paralysis, he
is in the habit of steadying himself, while on the road,
by keeping hold of the crupper of the saddle, or of the
tail of the ass. During a recent winter which was
more than usually severe, while on a journey near Rule
Water, the old man and the ass were suddenly immersed
in a wreath of snow, which had filled up a hollow in the
road. There they lay, far from help, and ready to
perish; till, at last, the poor ass, after some severe
struggles, extricated itself, and got upon safe ground.
But his master was still in the snow; so, after considering the matter for a while, the creature returned, forced
his way to his master, and then placed himself in such a
position as to give the poor pedlar a firm grasp of his
tail. The perishing man eagerly availed himself of this
help, grasped his ass's tail, and was immediately dragged
out by the faithful beast, till they both reached a place
of safety."

Mr. Thomas Fuller, in the *Naturalist*, says,—

" When on a visit to the neighbourhood of Marsh-
field, in Gloucestershire, while riding slowly along the
old Roman road, my attention was arrested by a jackass
standing close to the side of a high barred-gate leading
into a field, unmindful, apparently, of all around him,

the very personification of dulness. 'Can so stupid-
looking an animal,' thought I, 'possess the smallest
grain of sagacity?' As I watched him, his head moved.
By turning it sideways, he forced it, with some difficulty,
between the bars ; then, turning it to its natural position, the poor beast seemed in danger of throttling
between the bars. 'Surely,' I thought, 'the creature will
be strangled'; and I was about to dismount and to go
to its relief : when, lo ! after a minute's pause, the ass
dextrously lifted the gate over the latch, and pushed it
forward ; then released his head in the same way that he
had introduced it ; and at once walked boldly into the
field, where he could pick and choose for himself."

A correspondent of the *Animal World* writes :—

" The following touching anecdote of a donkey came
lately under my notice, and I think it fully illustrates the
strong affection of which this much-abused animal is
capable. The donkey in question was the property of a
gentleman residing in the neighbourhood of Edinburgh.
She had been in his possession for a good many
years, and was under the care of his gatekeeper, named
John. But 'Jenny's' long service at last came to an
end. Her master left his home to go abroad, and she
was sold to a lady. But what was the matter with
'Jenny'? She had a comfortable stable, and plenty
of food to eat, but that was left untasted, and she was

cross and irritable to a great degree; indeed, so much so that her new mistress's servants could not manage her, and were kicked if they dared approach her. What was to be done? It was better to get rid of such an animal; so a message was sent to John, to come and take away his old favourite. John obeyed the summons, and was conducted to the stable—the servants gathering round to see what sort of reception 'Jenny' would give him. As he approached the stable-door, he cried out, 'What's wrang wi' ye, Jenny?' At once the faithful creature, recognising the voice of her old friend, rushed out, and thrusting her head under John's arm, stood very firmly, declaring plainly by her attitude, 'I have found you now, and I will not leave you.' The old man was much affected, and could not keep from tears; and as he looked at 'Jenny's' face he saw that the skin of it was injured by the tears she had shed during her separation from him. · The mystery of her crossness and obstinacy in refusing food was thus explained—it had just been grief that was 'wrang' with poor 'Jenny.' I am happy to say she was given as a present to John, who led her back to his home, supplying her with a meal on the way, for she was weak from her long fasting.''

The Rev. Cæsar Otway says,—

" A lady of my acquaintance was walking along the road, near her house, one day, when she met a party of tinkers going towards Connaught; and one tall, bold-looking fellow, in all the excitement of intoxication, was belabouring his poor ass with a huge cudgel. She could not stand this cruelty. She spoke to the man : she might as well have expostulated with Beelzebub. She then tried more potent means, and asked if he would sell the creature ? He named three times its value, and was ready at that price, he said, to part with it. She gave it; and the beast, now her own, was set at liberty, and sent to graze in her paddock. It was not long in recovering its spirits and its good looks; but before any very long time had passed, the ass was stolen—it was surmised, by the very man who sold it ; and now, for three years, there were no tidings to be heard of the poor donkey. At last, one day, as his mistress was taking her usual walk, she met a man urging along an ass, straining and bending under a heavily-laden cart. As she drew near there was a sudden change in the demeanour of the ass. Its ears were raised, and its head also ; it even tried to break into a trot, and came and laid its nose on the shoulder of the lady, its kind mistress, with the plainest signs of joy. Again she had to purchase it, and so it went back to her paddock.''

Froebel, in his '' Travels in Central America,'' remarks that—

'' One of the most striking characteristics of the mule

is his aversion to the ass ; and, on the other hand, the pride he takes in his relationship to the horse. If an ass, at any time urged by an aspiring vanity, tries to join himself to a drove of mules, he will, in most cases, be kicked and lamed by his proud relations. On the other hand, the horse always takes a distinguished position when among mules. They crowd round him, and follow his movements, and show jealousy one of another—each striving to stand nearest to their distinguished relative. This instinct is made useful in keeping together a drove of mules on a journey. A mare is placed at their head, and this animal is led by a cord, and the whole drove is thus kept under control. The man who leads the mare is instructed, in case of an attack of the Indians, to leap upon the back of this animal, and to take refuge in the waggon-encampment, whither the drove is sure to follow him.''

III.

IN THE FARM-YARD.

PASSING from the stable to the farm-yard, we come to a variety of animals of whom, too often, we know little, —mainly because we usually pass them by in silence. Like the horse, however, they have faculties which would often excite our astonishment, if we paid any attention to the daily life of these creatures. We will give a few instances of this kind, beginning with some of those animals which are generally found both in the farm-yard and in the field.

Mr. Kenway, of Edgbaston, writes to the *Animal World* as follows :—

" I was walking with my father and a younger brother many years since in the neighbourhood of Bridport, in Dorsetshire, my native town, through fields leading to Hyde Farm, where about thirty or forty cows were at that time kept. A carriage-road led to the farm, and through a large field, rather steep, to the field in which the cows grazed. As we passed through the large gate

THE LEADER OF THE HERD.

(like a turnpike-gate) at the bottom of this field, we observed a bull at the top of the hill making his way through the hedge at the top; and fearing some harm, our father hurried us to a gate a little higher up, over which we climbed, and waited on the safe side to watch what the gentleman was after. To our great astonishment and delight, he passed quietly by us to the bottom of the hill, and with the greatest deliberation put his horns under the bars of the gate, raised it off its hinges, and carried it to the side of the road, where he laid it down, quite out of the way. Having done this, he began to bellow with all his might, and in a few minutes we saw first one cow come through the gap that he had made in the hedge, and then another, and another, until the whole herd were passed through—all streaming down the hill in the greatest excitement, kicking up their heels, and throwing about their heads and tails in the most ludicrous manner imaginable, to our great amusement. The gallant gentleman stood beside the gate until he had introduced his friends into a beautiful pasture, just ready for the scythe ; and not until he had seen all fairly enjoying the sweet repast did he attempt to partake of it himself."

Professor Bell says that a lady, a near relative of his own, witnessed the following incident :—

"A cow which was feeding tranquilly in a pasture,

the gate of which was open to the **road**, was much annoyed by a mischievous boy, who amused himself by throwing stones at her.   She bore with the persecution for some time; but at last she went up to him, hooked her horns into his clothes, and lifting him from the ground, carried him out of the field, and laid him down in the road.   She then returned quietly to her pasture, leaving him sufficiently frightened.''

" While playing at cricket," says Mr. Morris, "in a field belonging to J. Hayton Ireland, Esq., we were very **much** amused by watching his cow, when engaged in **slaking** her thirst.   As there is no pond near, **a** pump **had been** placed in the field, from which a stone basin below is usually kept filled for the horses and cow. But she, being nice in her taste, prefers to pump the water for herself, catching it, as it falls, with her tongue. The shape of her horns, bending **downwards,** enables her to work the pump with great **facility,** and we have often seen her **do it.''**

Mr. Otway tells of his own experience,—

"I am in the habit every year of buying two or three Kerry cows.   They are the kindest little creatures in the world, and I generally pick out those I consider to have good countenances.   Last year I was lucky in the three I bought; they soon became great pets.   They meet

me every morning at the gate of the pasture, expecting
to be spoken to. One in particular, a quaint little lassie,
used to put her nose into my pocket like a dog, to look
for a piece of bread, or potato. Well, there was a
swing in this field; and my Kerry lass, who was very
curious, seeing the girls swinging, thought, I suppose,
that she should like a swing herself. So one day,
about noon, a great lowing of cows was heard, and
some one who was at home went out to see what was
the matter. When he came to the gate, he saw two of
the Kerry cows in a great state of agitation; and they
followed him, lowing, to the further end of the field,
where he found the third, entangled in the swing,
caught by her head and horns, and in danger of being
strangled by her efforts to get out of the ropes. He
soon extricated and set her at liberty, and the alarm of
the other cows at once ceased."

A farmer living at Caversham bought two pigs at
Reading market, which were carried to his house in a
sack, and let loose in his farm-yard on the banks of the
Thames. The next morning they were not to be found,
and on inquiry being made, a person gave information
that two pigs had been seen swimming across the water.
Subsequently they were seen trotting along the Pang-
bourne Road, and in one place, where two roads met,
putting their noses together as if consulting. At last

they got back to the place from which they had been
taken, having travelled about nine miles by cross roads.
They were taken back to Caversham, but seized the first
opportunity to escape again, and made their way to
their first home with the same success. The strangest
part of the proceeding was, that they swam across the
river, keeping close together, in the exact line towards
their old master's house—of which direction, since they
had been carried in a sack, it is difficult to imagine how
they could have had any conception.

The next story has something lamentable about it.
It tells us of a poor pig who was certainly ill-used.
After being brought up in a drawing-room, it was all
at once consigned to a sty! The consequence was not
surprising: poor piggy died of a broken heart.

"A lady," says Mr. Morris, "took compassion on a
poor little pig, which was more than the mother could
find sustenance for, and determined to bring it up by
hand. It was petted, and so trained that in its baby-
hood it was a fit inmate for the drawing-room. It would
sit up like a dog, and beg for bread and butter, and
comfort itself like any other household favourite. But,
unfortunately, it grew up a fine pig, and one far too
large for the inside of any house. Its patroness was
perplexed what to do. At last it was resolved that
piggy should be handed over to a neighbouring farmer,

THE FARMYARD.

a very kind man, who promised to take care of it, and to save it from being made into pork. And as its mistress was going away for a week, she determined to take that opportunity for making the transfer. Wishing to treat piggy with especial indulgence, the farmer put it into a small paddock; but a way of breaking out was soon found, and the pig got back to its former quarters. Here, however, its mistress was 'from home,' and there was no one to take piggy's part. So it was taken back, and for security shut up in a sty. After the lapse of some days, its mistress returned, and soon after walked over to the farm-house. 'How is "Betty" (the pig's name) getting on?' she asked. 'Well, that's just what I was coming to you about,' said the farmer; 'my missis is uncommon vexed about that there pig that is lying dead out yonder. We were forced to shut her up, and then she wouldn't take her food, though we tried to tempt her with all manner of tit-bits; so she grew leaner and leaner, and looked so miserable, that it do seem a good thing she's gone; for I believe she'd never have been of any good to anybody.'"

A seafaring lad named Murphy, who lived with his parents near the Croft, at Hastings, and who was very fond of rearing dumb animals, died after a short illness. Among other pets, he had brought up a young goat, who was frequently seen in St. Clement's Upper Burial-ground. By some means this animal seems to have become conscious of the death of its friend and master. It seemed in great distress, and butted against the door, as if determined to effect an entrance. The attention of the relatives was excited, and at last they admitted the animal, who at once made its way to the room where the body lay, leaped upon the coffin, bleating loudly, and licking the poor boy's face.

Mr. Otway gives the following story :—

"At some flour-mills, near Clonmel, there was a goose, which, by some accident, was left solitary, without mate or offspring. Soon after, the miller's wife had put a number of duck's eggs under a hen, and, as usual, the ducklings, when hatched, ran eagerly to the water; and the hen, also as usual, stood clucking in great alarm on the pond-side. Just then, up came the goose, and cried out, in goose-language, 'Leave them to me; I'll take care of them.' She swam up and down with them, and when they were tired she brought them back to the hen, their foster-mother. The next day, down ran the ducklings to the pond, and the hen after them; and there was the goose, ready to take charge of them. Now, whether the goose suggested the thought or not, so it was, that when the ducklings took the water, and the goose lay close by to receive them, the hen, full of anxiety for their safety, jumped off the bank on to the goose's back, and there sat—the ducklings swimming, and the goose and hen following them up and down the pond. And this, once begun, was continued for weeks, until the young ducks began to be independent both of the goose and of their anxious foster-mother."

Captain Darke, of Plaistow, wrote in *Land and Water,*—

"About two years ago I bought a wild gander in Leadenhall market. I took it home and turned it loose among the other water-fowl; but he would not associate with them at all. He made friends with a cow at first; but after a few weeks he left her and took to a large Newfoundland dog, and this dog he never left, night or day. He ate and slept with the dog. He would fly at any one who interfered with the dog; and when I let the dog loose, the gander would run or fly close to him. If by any chance he lost sight of his friend, he would call out and run everywhere till he found him. About three weeks ago the dog sickened and became so weak with the distemper that he could scarcely stand or move. The gander remained by him, would not be driven away, refused his food, and when the dog died, the gander, in a few days, died also."

Mr. Bennett, in his "Wanderings in New South Wales," narrates the following history:—

"A drake was stolen one night, with some other
birds, from Mr. Beale's aviary. The beautiful male
was alone taken, and the poor duck was left behind.
The morning following the loss of her husband, the
female was seen in a most disconsolate condition;
brooding in secret sorrow, she remained in a retired
part of the aviary, pondering over the terrible loss she
had just sustained. While she was thus lost in grief,
a gay, prim drake, who had not long before lost his dear
duck, trimmed his beautiful feathers, appeared quite
handsome, and pitying the forlorn condition of the
bereaved one, sidled up to her and offered her his
protection. She, however, could not thus be comforted; she repulsed all his offers. Her usual avocations were neglected; her food almost forsaken; and no
comfort seemed to be found in sympathy or kindness.
Meanwhile, search and inquiry was made for the stolen
drake. Many days passed before a clue was obtained;
and then the bird could only be got back again by the
payment of four dollars. This was given, and he was
then brought back to the aviary in one of the usual
cane cages. As soon as the bird recognised his former
home, he showed his joy by quacking vehemently and
flapping his wings. Three weeks had passed since he
had been carried off; but no sooner did the forlorn duck
hear the voice of her lost partner than she quacked,
even to screaming, with ecstasy, and flew as far as she

could to meet him. So soon as the cage was opened the drake immediately ran into the aviary, and the pair were again united. They quacked and quacked, crossed their necks, and bathed together, and doubtless related to each other all their past hopes and fears. One word more, with reference to the unfortunate widower who had offered consolation to the bereaved duck, when her partner was taken from her. How the information was conveyed from the one to the other cannot be known; but certain it is that on the very next morning the returned drake attacked the neighbour who had tried to take his place, pecked his eyes out, and inflicted so many injuries as to cause his death in a few days."

Mrs. Lee describes an artful trick which a magpie, kept in the family of Mr. Ranson, was observed to play. A toll-gate stood near, and he was fond of watching the movements of the toll-keeper's wife. When he observed her to be employed in making pastry, he would perch upon the gate, and would suddenly cry out, "Gate ahoy!" If the husband were absent, as was frequently the case, the wife would hurry out to open the gate. Magpie would then slip in, and snatch a bill-full of the pie-crust—chattering over it, on the roof of the gate-house, with the greatest glee.

"On recently visiting the Zoological Gardens in the

Regent's Park," says Mr. Morris, " I observed a little incident which may be worth mention. A large white cockatoo and a much smaller green parrot inhabited the same cage. When I gave them a nut, the parrot took it, but instead of endeavouring to crack it, immediately handed it over to the cockatoo, whose more powerful mandibles at once mastered it, and the contents were fairly divided between the two."

"A person of my acquaintance," says Mr. Morris, "was very fond of pets, and had a number of rabbits, guinea-pigs, and other creatures, who were kept on a large grass-plat. Among these animals a fine rose-crested cockatoo used to wander, fearlessly, but without doing any injury to the rest of the creatures. One day my friend took a fancy to a fine white Angola rabbit, which he bought, and placed with the others on the grass-plat. This new animal attracted the attention of the cockatoo, who straightway walked up to the rabbit. The latter did not seem at all afraid, thinking, perhaps, that as they were both white, there must be some affinity between them. But when the cockatoo had got close to the rabbit, he put his beak to the other's ear, and called out, 'Who are you?' The consternation of the rabbit was most amusing; he bounded off in an instant to the farthest extremity of the enclosure."

The Rev. Edward Spooner writes,—

"I have been lately visiting a friend in Staffordshire, who owns a large grey cockatoo. 'Poll' is a most communicative bird, and a great friend of the family. On a fine day she generally passes several hours in the back-yard, outside her cage; for, though unchained, she rarely leaves the house. She is on good terms with all the yard-dogs, the house cats, and the poultry; but if a strange dog or cat enters the yard, she flies at him at once with a tremendous scream. At night she sleeps in the kitchen, where her usual companions are three cats. One morning the kitchen-maid went down-stairs early, and before she entered the kitchen she heard 'Poll' talking loudly. On opening the door, she found 'Poll' seated on the dresser with a large piece of bread in her claw. Round her, on the floor, were the three cats, and a chicken which had lately taken refuge there. With strict impartiality, the bird was breaking off pieces of the bread, and dropping them to her pensioners in turn, who received the dole without squabbling, and with gratitude, listening all the time to all the words in her vocabulary, which were poured forth in rapid succession."

A gentleman who was lodging at the "Red Lion Inn," at Hungerford, says,—

"Coming into the inn-yard, my chaise ran over and bruised the leg of a favourite Newfoundland dog; and

while we were examining the injury, 'Ralph,' the raven,
looked on also, and was evidently making his remarks
on what was doing ; for the minute the dog was tied up,
'Ralph' not only visited him, but brought him bones,
and attended him with many marks of kindness. I
remarked this to the hostler, who told me that the bird
had been brought up with a dog, and that the affection
between the two was mutual, all the neighbours being
witnesses of the many acts of kindness done by the one
to the other. The dog once had his leg broken, and,
during a long confinement, the raven waited on him
continually, brought him food, and scarcely ever left
him alone. One night, by accident, the stable-door was
shut while 'Ralph' was outside, and thus the bird was
shut out from his friend's society; but in the morning
the hostler found the door so pecked away that in
another hour or two the bird would have forced an
entrance."

Dr. Stanley tells us of a turkey-cock, who was so
unremitting in his attentions to his mate that his
solicitude was respected, and when she was sitting
on her eggs, he was allowed to be with her. Soon,
it was perceived that he had claimed a part of the
duty, and had taken some of the eggs from her,
which he covered with his own body. The servant
who attended to the yard doubted the propriety of

this, and put the eggs back again under the hen.
No sooner, however, had her mate the opportunity,
than he again took possession of some of them. The
master of the house took it into his head to let the
male turkey have his own way; and he ordered a

nest to be prepared, with as many eggs as the bird
could cover. The turkey seemed highly pleased, and
sat with great patience, hardly quitting them to get
his food. After a while, from the two nests twenty-
eight young ones appeared, and the male bird, when

he saw this large assemblage of young turkeys, looked in no small measure perplexed.

" A Canada goose," says Dr. Stanley, " was observed to associate itself with a house-dog, and would never quit the kennel, except for the purpose of feeding—when it would return again immediately. It always sat with the dog, but never presumed to go into the kennel, except in rainy weather. When the dog barked, the goose would cackle, and run at the person at whom the dog barked, trying to bite his heels. **When the dog** went out of the yard, and ran into the **village,** the goose always accompanied him. This extraordinary attachment was said to have originated in his having once saved **her** from a fox. When the dog became ill, the goose never quitted him, day or night ; and she would have been starved had not a pan of corn been placed near the dog's kennel. During his illness she generally sat close by him, not suffering any one to approach, except those who brought her own or the dog's food. When the dog died, she still kept her place by the kennel; but unfortunately a new and strange dog was introduced, and when she approached him, as she had done her former friend, he seized her by the neck, and she died in his grasp."

Dr. Stanley tells us of a cock which became the

COCKS FIGHTING.

terror of the poultry-yard, and which was so pugnacious that if his owner passed him in the yard without
offering him some food, the cock would attack him, and
even if driven off, would return to the attack.

Mrs. Lee says: "On one occasion I saw a cock pursue
a hen round the poultry-yard, and when he had caught
her, he took a worm from her, and gave it to another

hen, who stood by, waiting. I came, however, to the
opinion that the hen who was punished had stolen the
worm from the other, and that the cock was only doing
justice between them."

A warm friendship was formed for a large otter-dog,
by a raven, which was kept in the same yard. At

first the bird merely hopped about the dog's kennel, and picked up occasionally a scrap from the dog's pan, when he had finished his meal. By degrees the acquaintance improved, and the bird became a constant guest at meal-times, taking up his position on the edge of the dish, and helping himself to the best bits. Often the bird would snatch up a piece of meat, almost from the very mouth of the dog, and running beyond the reach of his chain, would tantalize him with it; ending, however, generally, in a good-humoured surrender to his friend. This intimacy was terminated, at last, by a mischievous boy, who killed the poor raven by suddenly throwing a stone at it.

IV.

IN THE FIELD.

"Mr. Topham, of Armley," writes Mr. Morris, "has a shepherd's dog which was brought up with a pet lamb. When the lamb no longer required fostering care, it was sent to the flock ; and the young dog, having grown up also, had to do its duty in driving the sheep. But the two friends are friends still ; and the sheep, instead of running away from the dog, like the rest, walks quietly by his side, while he drives all the other sheep before him."

A gentleman travelling in his gig in a lonely part of the Highlands, was met by a ewe, who came up to him bleating. She cried louder as he drew near, and looked up to him as if to ask assistance. He stopped, went to her, and when she turned, he followed her. She led him to a little hill at some distance, where he found a lamb which had fallen down between two large stones, and lay crying and struggling to extricate itself. He pulled away one of the stones, and helped the lamb to get out ; and then the mother, in evident delight, poured forth her thanks in a long-continued bleat.

"One day in last April," says a writer in the *Magazine of Natural History*, "I observed a young lamb entangled among briars. It had struggled for liberty till it was nearly exhausted. The mother then attempted to release it, both with her head and her feet, but all in vain. Finding the task too much for her, she turned round and ran away, baa-ing with all her might. She went across two or three fields, and through their hedges, till she came to a flock of sheep. In about five minutes she returned, bringing with her a large ram, whose horns gave just the help that was wanted. They went together to the poor lamb, and the ram immediately set about the work of releasing it, dragging away the briars with his horns."

A similar narrative is given by Cuvier, who relates how a sheep applied, bleating, to a cow; and when the cow followed her, led her to a lamb which had fallen into a ditch, and lay feet uppermost, unable to release itself. The cow, by a careful use of its horns, lifted the lamb out of its hole, and placed it upon its feet again.

Mr. St. John says that he received from a most reliable informant the following relation :—

"Very early one morning, he saw a fox eyeing most wistfully a number of wild ducks feeding on the rushy end of a Highland lake. After due consideration, the fox, going to windward of the ducks, put afloat on the lake several bunches of dead rushes or grass, which floated down among the ducks without causing the least alarm. After watching the effects of this manœuvre for some time, Reynard, taking a good bunch of grass in his paws, launched himself into the water as quietly as possible, floating down, as the preceding bunches had done, towards the ducks. He took care to leave nothing but the tip of his nose and ears above water, and these were concealed among the grass. In this way he drifted among the ducks, and soon made booty of a fine mallard. When we remember what numbers of wild ducks, pigeons, hares, and other wild animals every fox contrives to catch, for his own support and that of his family, this story seems by no means incredible."

The fox will counterfeit death in a most extraordinary manner. It will sometimes allow itself to be taken by the brush or tail, and flung out upon a dunghill, and then, soon after, will awaken up and take to its heels, to the dismay of the bystanders. " On one occasion," says Mr. Mudie, " a man going to his work through furze-bushes, on a common, came upon a fox, lying stretched out at length under one of the bushes. He drew the creature out by the tail, and swung it to and fro, and then threw it on the ground; but not a sign of life did it show. He then took the fox by the tail, and swung it

over one shoulder, hanging his mattock over the other,
and so trudged along the high road. The fox had
shammed death to admiration ; but the mattock rendered his present position intolerable, for the point of
it came against his ribs. So, at last, he gave a **snap,**
or bite, which penetrated the flesh at the lower part **of**
the man's back, and forced him instantly to throw down
his load. At first he could not imagine what had
happened, but looking round, he saw the fox, which he
had imagined to be dead, making off for the nearest
thicket at the greatest possible speed."

At **Kilmorac, in** Inverness, the parochial minister was
a man of great taste and much hospitality. He kept a
good stock of poultry; but, as foxes were numerous, it
was needful to provide the fowls with a house for greater
safety. A visitor found on the breakfast-table, not only
salmon fresh and salmon preserved, but **new-laid eggs**
in daily abundance. One **morning the** servant, whose
duty it was to provide the latter, took the key, and the
usual basket, and repaired to the fowl-house, to bring in
the usual supply. But when she opened the door, terrible
was the scene. Blood was here and there, and dead
hens lay on every side. In the midst lay a full-grown
fox, apparently **as** dead as the hens. The servant
stared, and wondered, **and** supposed that the fox must
have eaten himself to death. After some exclamations

and evil-speaking, she took the beast up by the tail, and
flung him out of the window, on to the dungheap outside. But what was her astonishment to see the creature, so soon as he alighted on the heap, spring up, alive
and vigorous, and scour away for the neighbouring
woods!

A fox, partly tamed, was kept fastened by a chain to
a post in a court-yard, and was chiefly fed on boiled
potatoes. Many fowls also were kept in the same yard,
but they had sense enough to keep out of reach of the
fox. He was, however, too cunning for them at last.
He bruised and scattered some of the potatoes given
him for dinner as far to the extremity of his circle as the
chain would allow, and then, retiring in an opposite
direction, he laid himself down and feigned to be fast
asleep. After awhile his stratagem succeeded; some
of the fowls, seeing their enemy entirely dormant,
ventured near, and began to pick up pieces of the
potatoes. But the fox was silently watching them, and
when he saw one of them within his reach, he made a
spring, and instantly got a fowl for his dinner.

Captain Lyon writes that he had an Arctic fox, whose
tameness was so remarkable that he did not like to have
him killed; but kept him on deck, in a small hutch,
with a length of chain which allowed him to take some

exercise. But the fox found that people were glad to get hold of this chain, and with it to drag him out of his hutch, to gratify their curiosity. To prevent this, he adopted the precaution of drawing all the chain after him into the hutch, so that no one could get hold of it without putting his fingers within reach of the animal's teeth.

Captain Lyon also remarks of these Arctic foxes, that they can so modulate their bark as to make the hearer think that the fox is at a distance, when all the while it is lying at his feet. They have also the power of decoying other animals, by imitating their voices. He once observed a fox prowling about on a hill-side, and imitating the cry of a wild goose.

Mr. St. John, in his "Sports of the Highlands," describes a fox whose manœuvres he had himself watched :—

"Just after daylight I saw him come quietly along. He looked with great care over the turf-wall into the field, and seemed to long very much to get hold of some of the hares which were feeding within it ; but apparently knew that he had no chance of catching one by dint of running. He seemed, therefore, to form a plan. With great care and in silence he scraped a small hollow in the ground, just where the walk seemed to be most frequented ; he threw up the sand as a

screen, stopping often to take a view of the field.
When he had prepared his 'rifle-pit,' he laid himself
down, in the best position for springing on his prey,
and remained quite motionless, except an occasional
look towards the feeding hares. When the sun began
to rise, the hares thought it time to leave the field for
the plantation; and his ambush was laid with a view to
this. Three passed by him, but he stirred not: they
were scarcely within his reach. Two more came
directly towards him; and now, with the quickness of
lightning, he had sprung upon one, and killed her
instantly. He took possession, and was carrying her
off, when my rifle-ball stopped his course."

Mr. Jesse says that, when a fox is troubled with fleas,
he will go into the water,—not suddenly, but at first to a
small depth, the water only covering the lower part of
his body. The fleas then creep upwards, and rise to his
back. Presently he will go deeper, till the water covers
his body, and only his head is above it. The fleas are
then driven forward, till they get together in a swarm
on his head and nose. At last he will lay in breath
enough, and plunge his whole head under water, washing
the fleas entirely off. A friend told Mr. Jesse he saw a
fox doing this in a lake in Italy.

Mr. Garratt describes another sort of sagacity. A

farm-servant was ploughing a small field for wheat, in Ireland, and was surprised to see a fox pacing slowly along in the furrow just before the plough. Soon he heard the cry of the hounds; but, turning round, he saw the whole pack brought to a stand at the other end of the field, just where the fox must have entered on the land under the plough. Somehow the animal must have thought of this way of breaking the scent.

In Leicestershire, an old fox, often hunted, was always lost at a particular place; after coming to which, the hounds always lost the scent. It was at last discovered that he jumped upon a close-clipped hedge, ran along the top of it, and then crept into the hollow of an old pollard-tree, where he lay snugly concealed till the hounds were baffled and drawn off.

An old man was walking along the banks of a river, when he observed a badger moving leisurely along the ledge of a rock on the opposite bank. After a while a fox came running up, followed the badger's track for some distance, and then leaped into the water. In a few minutes a pack of hounds came up in full cry. Seeing no fox, they fell upon the badger, who, in a minute or two, was torn to pieces. Meanwhile the fox was going down the stream, breaking the scent; and at last crept into a hole by the river side.

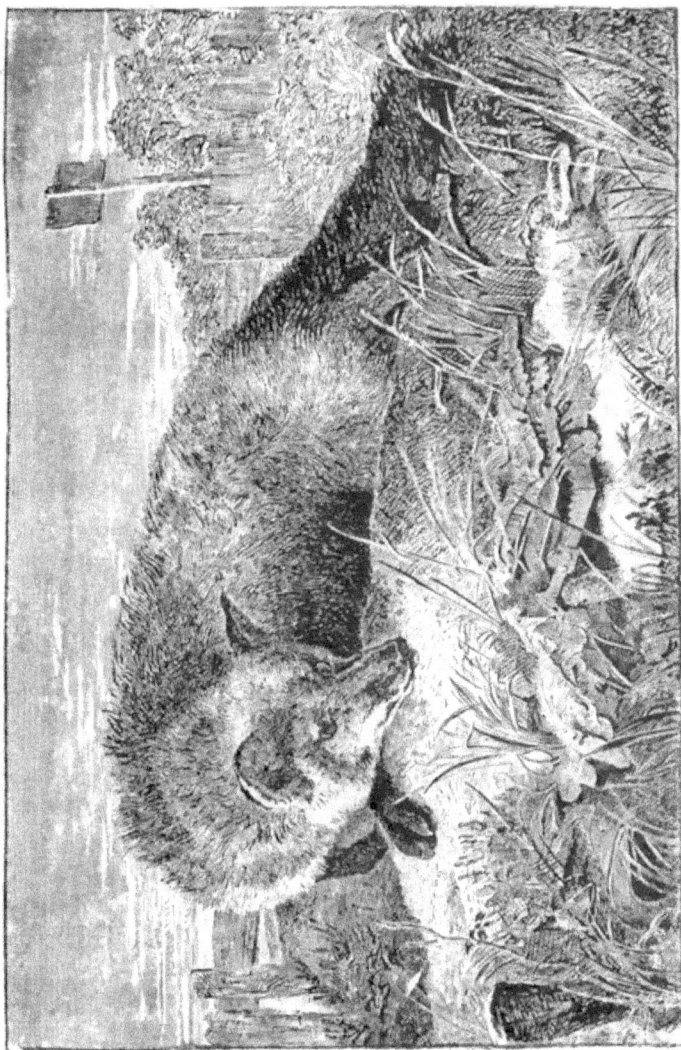

NOT CAUGHT YET.

Mr. Moffat, in the *Naturalist*, thus describes a fox which he himself had tamed :—

" For several days (being very young) it was very discontented with its new situation, and kept almost constantly calling out in a sort of quick, yelping bark. But it lapped milk readily, and soon became reconciled, and so tame, that until it was twelve months old it had its liberty about the house like a dog, which in many points of character it closely resembled. When more than half-grown, it used to follow me about the garden and village, and frequently made little excursions amongst the cottages by itself, popping into the houses to investigate their larders, to the no small terror of the old women: and frequently have I seen them endeavouring to drive him out with a broom ; but Reynard would contrive, somehow or other, still to pop past them with the most impudent effrontery, and not to leave the house, in spite of invective, till it suited him to do so. Every night, while it remained an inmate of the dwelling-house, it slept upon a mat at the foot of the staircase, and as each member of the family came down in the morning it used to meet them half-way, and to express its joy by leaping against them, and uttering a sort of hoarse scream, much after the manner of an affectionate dog, and fanning its tail in the same way ; only the expression of its satisfaction seemed more extravagant. When nearly full-grown, Reynard's natu-

ral predilection for poultry manifested itself one day,
in his being met by the servant carrying in his mouth a
turkey-hen, which he had taken from her nest in the
yard. For this act of felony he was confined, for the
future, by a chain on a grass-plat, behind the house,
with a dog-coop for shelter. But he preferred a hole of
his own digging, in which he generally lay. His confidence and affection for the family was not impaired by
this confinement; but his disposition towards strangers
was much altered. He seemed now distrustful and
suspicious of every one with whom he was not acquainted; he watched them with intense attention, and
on their approach ran into his hole, from which it
required much force to drag him. Still, if any of ourselves went towards him, he would try to meet us at the
stretch of his chain, uttering a peculiar scream indica-
tive of his satisfaction. I have even seen him leave his
meal, to welcome my youngest sister, who was an
especial favourite. And, what was most singular in an
animal naturally so wild,—he would allow me to open
his mouth, to place my fingers in it, even down to the
throat, as often as I pleased, without once attempting
to bite. I could also, at any time, take him up in my
arms: a liberty I should not like to take even with the
quietest dog."

"A gentleman," says Mr. Youatt, "near Laggan,

in Scotland, had a bull which grazed with the cows in the open meadows. As fences were little known in those parts, a boy was kept to watch lest the cattle should trespass on the neighbouring fields, and injure the corn. The boy was idle and drowsy, and often fell asleep; then the cattle trespassed, and he was punished. In his turn, he revenged himself on the cows, whom he punished for their transgression with an unsparing hand. The bull seemed to observe this, and to understand it. He began to keep the cows within bounds, and to punish them if they strayed over the boundary. He himself never entered the forbidden ground, and if he saw them approach it he placed himself in the way in a threatening attitude. At last his honesty and watchfulness became so obvious, that the boy was withdrawn and set about other work, and the duty was left wholly to the bull."

The sheep is usually regarded as a creature of small intelligence and little sympathy; but this notion chiefly arises from our disregard and want of observation. One who knew them well, said, " The marked character of the sheep is that of natural affection, of which it possesses a great share." Mr. Youatt (writing some years since) said, " At the present moment there is, in the Regent's Park, a poor sheep with a very bad foot-rot. Crawling along the pasture on its knees, it with

difficulty contrives to procure for itself subsistence;
and the pain it suffers when compelled to get on its feet
is evidently very great. At a little distance a companion, watching over it, is always to be seen ; and that
companion is always the same animal."

Mr. Hall, in his travels in Scotland, says, "I was
one day climbing the mountain of Belrennis. On
reaching the top, I found myself wrapped in a mist,
beyond which I could only see a few yards. I waited
in hope that the mist would disperse, so that I might
see my way. Cold and hungry, and walking about to
keep myself warm, I perceived something uncommon
at a little distance, and approached it, not without some
fear. I soon found it to be a phalanx of sheep, drawn
up on the top of the hill, and ready to defend themselves
against any attack. In the centre of the line was a large
ram, with a black forehead and a tremendous pair of
horns. The weaker ones of the flock were in the rear.
Not one of them was eating, but all looking sternly upon
me. Seeing them in this warlike array, I began to feel
a little alarm, and slowly retired. Had I approached
them, I believe that they would have made a simultane-
ous rush upon me."

The he-goat is a most powerful animal for its size,
and sometimes presumes upon its strength in a most

intolerable manner. William Howitt tells us of one who became the tyrant of the village. When it left its customary yard, and marched down the village street, the women and children all kept within doors. On one of these occasions a poor old woman, who was creeping down the road, was either too weak or too deaf to get out of his way. Without the least hesitation the goat knocked her down. The whole population was greatly excited, and all the men agreed that " it was a shame"; but no one liked to " tackle the bully." At last one man produced a heavy pole, or bed-post, and with it made a charge, and the blow resounded on the goat's head—only to add to his wrath. He made one spring upon his assailant, dashed the pole out of his hand, and laid him flat on his back—the victor standing in triumph over him. All was terror and confusion ; when, happily, two men came in search of the runaway. One of them knew how to deal with him. Taking hold of his beard with a firm grasp, the hero was led off, complaining greatly.

" Two robbers," says Mr. Morris, "took a pig, weighing fourteen stone, out of its sty, and drove it along a lane leading to Rotherham. On coming to a lonely path across the fields, they thought it would be prudent to kill the animal in this quiet, out-of-the-way place, where its cries would scarcely be heard. One of them

took out a knife, and began cutting the pig's throat.
But the creature, thus alarmed, struggled violently, and
managed to escape out of his hands, running squealing
into the next field, with a gash in its throat. The men
ran after it, but found in the field a bull grazing; and
this animal at once took the pig under his protection.
As the robbers came towards the pig, the bull ran at
them, and they only narrowly escaped being tossed into
the air. Rushing back through the fence, they lingered
outside, hoping for another chance; but piggy kept
close to his new friend, and they had to go home at last
without him. On their apprehension, some time afterwards, they confessed; and the pig's owner thus had
the mystery cleared up—how his pig came to be found
in a distant field, with its throat partly cut, and keeping
close company with a bull."

The deer tribe are, generally, very fond of music.
Mr. Playford says, "Some years since, I met on the
road near Royston a herd of about twenty bucks, who
were following a bagpipe and violin. While the music
played, they followed it; when it ceased, they stood
still; and in this way they were brought out of Yorkshire as far as Hampton Court."

"The deer," says Mr. Wood, "are very gentle,
except at certain times of the year, when it is dangerous

THE RIVALS.

to come near them; for the bucks are ready to fight anything that comes in their way. It was at such a time that a gentleman, fond of sketching, had ventured into a park, thoughtless of the danger, and was engaged in his work, when a deer saw him, and instantly charged down upon him. He was only too happy to escape by sacrificing everything, and swinging himself up into a tree, one of the branches of which happily hung near him."

Mr. St. John adds, "I once saw, in a public garden near Brighton, a beautiful but small roebuck fastened with a chain which seemed heavy enough for an elephant. I pitied the poor animal; but the keeper told me that this heavy bondage was necessary, for only a few days before he had killed a boy of twelve years old, —inflicting more than fifty wounds in a very few moments."

A nobleman, towards the end of the last century, took a fancy to tame four stags, and to make them run in a light carriage. Naturally this freak excited great attention, and the owner of this novel equipage was fond of exhibiting himself in it. But, unfortunately, he was driving, one day, near Newmarket, when a pack of hounds, out for the day, crossed the road just behind him. The stags instinctively set off at their utmost speed, and the dogs followed, nothing loth.   The pace grew more and more furious, and the driver began to fear for what might happen; when, happily, an inn in Newmarket came in view, where he was in the habit of stopping.   The whole *cortège* dashed into the yard, and the hostlers and attendants were just able to keep the dogs from pouring in after them.

V.

IN THE WOOD.

A GENTLEMAN of Worthing writes,—

"An instance of squirrel sagacity has lately been re-lated to me by an eye-witness. A tame squirrel, having climbed up the bell-rope, threw down from the end of the mantelshelf an imitation egg and egg-cup in one. He immediately descended, and tried, having secured it in his paws, to re-ascend and put it back into its place again. Not being able to do this, he again climbed up the rope; but, not satisfied, he came down again, and now, having managed to secure the ornament in some way which left his paws at liberty, he again ascended the rope, and replaced the thing on the mantel-shelf."

"Owls," says Bishop Stanley, "have been noticed for extraordinary attachment to their young. One day, in the month of July, a young bird, having strayed or fallen from the nest, was caught by some servants. It was shut up in a large hen-coop; and the next morning a fine young partridge was found lying dead before the door of the coop. Night after night did the old birds pay a similar visit, for the same purpose. The provision thus brought was usually young birds, newly killed. A watch was set, several nights, to observe how this supply was

brought; but the parent-owls were very quick-sighted, and observed any moment when the watch was relaxed, and at those moments the food was placed before the coop."

Mr. Couch adds,—

" A brown owl had long been in possession of a convenient hole in a hollow tree; and in this nest had
deposited its young, season after season, but only to be
robbed by some of the people of the farm, who had
watched and discovered its retreat. At last, in the very
presence of the mother, a plunderer was seen ascending
the tree,—when the parent-bird, aware of the danger,
grasped her young one in her claws, and bore it away;
and never after that was the nest placed in that perilous
position."

The *Dumfries Courier* vouches for the following
story :—

" While Mr. Charles Newall, granite-hewer, in Dal-
beattie, was plying his vocation in Craignair quarry, his
attention was suddenly arrested by cries strongly indi-
cative of distress, proceeding from one or more of the
feathered denizens of the wood. On throwing down his
tools, and hurrying to the spot whence the sounds proceeded, he discovered a robin, apparently in a state of the
greatest agitation, whose movements quickly showed him
the true cause of alarm. An adder, twenty inches long
and one inch in circumference, had managed to drag
itself up the face of the quarry, and was at that moment
in the act of protruding its ugly head over the edge of

a nest built among the stumps of brush wood, which
contained poor Mother Robin's infant offspring. She
was engaged, when Mr. Newall first got his **eye** on her,
in alternately dashing down upon the intruder, darting
her beak into its forehead, and then rising on the wing **to**
the height of a yard or so above the scene of danger.
In a moment Mr. Newall had dislodged the aggressor;
but in doing so, two of the little birds were thrown out
of the nest—into which, however, they were speedily
returned. While Mr. Newall was engaged in killing the
adder, the joy of the parent-bird was so excessive that she
actually perched on the left arm of her benefactor, and
watched with intense delight every blow inflicted on her
merciless enemy; and when that enemy was dead, she
gratified her feelings by alighting on and pecking the
body with all her vigour. She then returned to her
nest, and, having ascertained that all was safe, she **flew**
to a neighbouring branch and sang her **song of triumph
and hymn of** thanksgiving."

A singular anecdote is related by Mrs. Lee of a pair
of goldfinches. They had built their nest on a small
branch of an olive tree; but it appeared that when
the brood was hatched, and began to grow, the weight
of the nest was too great for the bough on which it
depended. Something was necessary **to be** done. The
parent-birds found a small piece of string lying on the

walk; and with it they fastened the bough, with its dependent nest, to another and a stronger bough, which hung a little above it, **and** thus made all safe.

Lord **Brougham** gives us the following story, which, he says, **was** told by Dupont de Nemours, who adds that he himself witnessed the affair :—

" A swallow, at Paris, had entangled its foot in **a** noose of a cord attached to a pump in the *College des Quatres Nations*, and it soon found that its struggles **to** get free only drew the cord tighter, so as to make **its** escape more and more impossible. It fluttered and struggled till its strength was exhausted, and it could then do nothing but utter cries of distress and despair. These cries, however, brought together a large number of its companions—indeed, it seemed as **if** all the swallows of the Tuilleries and Pont Neuf were collected together at the spot. For a time they seemed to consult **as** to what could be done. At last a plan was decided **upon,** and they all began a rapid and continuous **flight,** swallow after swallow, every one darting its beak **at the** string **which** kept the foot in bondage. This assault went **on,** unceasingly, for about half an hour, and then the string gave way, having sustained a thousand pecks; and the prisoner, in a moment, found himself free. A loud chirping and chattering followed, of joy **and** congratulation, and then the assemblage

11

ended, and each bird betook himself to his own
affairs."

The following story is given by the Rev. T. O.
Morris :—

"Living in the City portion of the great Metropolis, I
observed, one afternoon, in the aperture generally left for
the cellar or kitchen window, an unfledged house-sparrow,
which had fallen down into this underground place, across
which was laid, obliquely, an iron bar, which extended
within a foot of the surface. The mother was at the top
of the opening, looking down with pity and alarm at the
condition of the child. Many and ingenious were the
attempts of both mother and child to raise the latter out
of its perilous position, but all of them proved unavailing.
I looked on with anxiety, lest the affair should end in the
mother's giving up the attempt as hopeless, and desert-
ing her child; but the event proved that there was no
ground to fear any such desertion. Several plans and
attempts were frustrated, one after the other ; but at last
the parent hit upon a new one. Flying away for a few
minutes, it at length returned with a stout straw in its
beak, and rested with it for a minute on the edge.
At once the little nestling, called by its mother by
divers chirps, climbed up the iron bar, till it reached the
highest point, and there received one end of the straw in
its beak, and was raised, to my astonishment, by the

mother, who held fast the other, till it reached the level ground."

A writer in *Science Gossip* says,—

"I had heard that swallows will 'mob' and put to flight a kestrel-hawk; but I was rather sceptical till lately, when I happened to have ocular demonstration. I had gone to see an old castle in the neighbourhood, which was built on the only hill for miles round, and was therefore certain to be the haunt of a pair or two of hawks. I kept my eyes open, and was soon re-warded by the appearance, on the brow of the hill, of a bird, which, by its graceful form and the hovering motion of its wings, I knew to be a kestrel. His active little enemies, the swallows, a flock of which were disporting themselves close by, were as quick to see him as I. They at once advanced to meet the intruder, and with the utmost audacity brushed past him in all directions, one from one quarter and one from another, wheeling and then returning,—the hawk making futile dashes now and again, but always too late to reach his nimble little opponents. At last, tired of the un-equal warfare, the hawk beat a hasty retreat. But he was not allowed to get off easily, but was followed by his victorious foes; and the apparent mystery was quite cleared up; for though he made off at his best speed, the swallows, with the utmost ease, fetched him up, passed him, wheeled round and met him, and then, taking another sharp turn, repassed him; repeating these manœuvres a dozen times or more."

Mr. Morris tells us that—

"On one occasion a pair of martins built their nest on an archway of the stables of Woburn Abbey, Bedfordshire, and as soon as they had completed it, a sparrow took possession of it. The martins tried several times to eject him, but they were unsuccessful. Nothing daunted, they flew off to get help, and in a **short** time returned with thirty or forty martins, who dragged the intruder out, took him to the grass-plot close by, and there fell upon him and killed him."

In another case, the same sort of friendly co-operation was given, not for revenge, but for a better purpose. Mr. Jesse says,—

"A swallow's nest, built **in a** corner of a window facing the north, and containing a brood of young ones, was so much softened by the beating of the rain upon it, that it finally gave way, fell, and was broken in the fall, leaving the young ones exposed to the weather. **The** owner of the house, compassionating the young creatures, ordered a covering of some **sort to** be thrown over them. As the storm subsided, many other swallows, interesting themselves in the parents' trouble, gathered

about the spot; and when, at last, they saw the covering removed, and the little ones safe, they showed every
sign of joy. While the parents fed their young, the
other swallows set to work, each one collecting clay and
other materials, and helping in the work, till they had
built over the brood a sort of arched canopy, and so
sheltered them from the weather. From the time it
took the whole flock to perform this, it was clear that
the young ones must have perished of cold and hunger
before the parents could have completed a fifth part of
the undertaking."

The Rev. Philip Skelton says,—

" I once saw a remarkable instance of the sense and
humour of the swallows, played off upon a cat which
had, on a very fine day, placed herself on the top of a
gate-post, as if in quiet contemplation,—when about a
dozen swallows, knowing her to be an enemy, took it
into their heads to tantalize her. One of these birds,
coming from behind, flew close to her ear, and she
made a snatch at it, but was too late. Another, in five
or six seconds, did the same, and she made the same
unsuccessful attempt to catch it. Then followed a third,
and a fourth, and all the rest; and every one, when it
passed, seemed to set up a laugh at the disappointed
enemy. They then formed a kind of circle in the air,
and flew round and round her for nearly an hour; till

at last pussy, tired of being made a butt of, jumped down and fled, as much huffed, I believe, as I had been diverted."

A similar community of thought and action is found among another large family of " winged things." Major Norgate, who had long been an observer, says,—

" The crows hold meetings for some reason or other. Two or three will begin cawing, and in a minute or two, forty or fifty others will come flying to the place by twos and threes, from every quarter. They then form a kind of ring round one crow, who appears to have been an offender against some of their rules; and they remain still for some minutes,—the culprit never attempting to escape. Then, all of a sudden, five or six of them will fall upon the prisoner, pecking him, and striking him with their wings. On one occasion, I saw the crow left dead on the spot; and on another, its wing was broken. Of course, as to why these crows are punished can only be a matter of surmise."

Mr. Wood confirms this statement, on the authority of a lady who, herself out of sight, was witness to one of these strange assemblies. A long confabulation was held, and then the bird who was condemned was nearly pecked to pieces, his mangled body being left on the ground.

Mr. Drew also, in *Science Gossip*, of October 1871, describes a similar trial, conviction, and execution, of which he was witness, at Nansladron, in Cornwall.

The following narrative comes from Germany:—

" A gentleman of property, residing in a mansion near Berlin, found that a pair of storks had built a nest on one of the large chimneys of his house. Curiosity led him to mount a ladder and to inspect it, and when he found in the nest one egg, in size much like that of the goose, the idea struck him that he would exchange it for one belonging to that bird. This was done, and the birds seemed to take no notice of the exchange. But when the egg, in due time, was hatched, the male stork soon perceived the difference, and flew round and round the nest, making loud cries, and at last disappeared, and did not return until two or three days had elapsed. Then, however, about the fourth morning, the people of the house were disturbed by loud cries; and, looking out, they perceived a large assemblage of storks, which they estimated at ' five hundred.' These assembled in one body, and an old bird seemed to be addressing the rest with great earnestness. When he had finished, another followed, and then a third; and, at last, the whole body simultaneously rose in the air, uttering dismal cries. The female, all this while, had been sitting on the nest, sheltering the young bird, and showing signs of alarm. All at once, the whole body of storks poured down upon her, knocked her out of the nest, and soon left her dead,—destroying also the poor gosling, and leaving not a vestige of the nest itself. They then separated, and soon disappeared, and the tragedy ended."

We return, for a moment, to the vicinity of our homes.

Bishop Stanley tells us that—

" A water-hen, observing a pheasant feed out of one of those boxes which open when the bird stands on the rail in front of the box, went and stood in the same place as soon as the pheasant left it. Finding that its weight was not sufficient to raise the lid of the box, it kept jumping on the rail to give additional impetus. But this not being sufficient, it presently flew away, and returned with another bird of its own species. The weight of the two proved sufficient, and they gained the reward of their sagacity."

Dr. Stanley describes a pair of Muscovy ducks that were landed at Holyhead from a Liverpool vessel. The male was sent to a gentleman's house, and put with other ducks; but to these he showed the greatest in-difference. He evidently pined for the loss of his mate. After a time she was brought to him, and let loose in

THE WIDOW.

the yard. At first he did not see her. But when, on turning his head, he caught a glimpse of her, he rushed to her with a joy which was quite affecting. Nothing after that would induce him to leave her; he laid his beak upon hers, he nestled his head under her wing, and often gazed at her with evident delight.

"The wild duck," says Sir William Jardine, "generally frequents some chosen piece of water or morass,

where the flock remains, making excursions, morning and evening, to various feeding-grounds. As twilight approaches they may be seen by the watcher, early in the night, coming from the daily resting-places. They at first fly round in circles, gradually lowering and surveying the ground ; but as the night advances, they fly straight to the spot and alight at once."

Our tame species comes from the mallard, or wild duck ; and they never lose their preference for marshy places and bogs.

No bird is an object of so much terror as the hawk. The alarm in a dove-cote when one is seen approaching, or the agony of fright into which a hen is thrown when a hawk hovers over the poultry yard, is well known to all dwellers in rural places. Mr. St. John tells us that he one day found a sparrow-hawk standing on a large pigeon on the drawing-room floor, and plucking it,—having entered by the open window in pursuit of the unfortunate bird, and having killed him in the room.

Dr. Stanley tells us that a gentleman, walking over his fields in Yorkshire, saw a small hawk attempting to fly off with some prey it had captured, but which proved too heavy for it. It was pursued by a hare, which, whenever it came within her reach, struck at it, and at last succeeded in knocking it down, when it relinquished its prey. This proved to be a young leveret, and the

AN INTRUDER.

hare was its mother. The little thing was wounded; but the mother's care might revive it.

From the wild-duck to the gull, the transition is easy.

"There is naturally less fear of man shown by gulls," says Mr. Morris, "than by most other birds. One can

scarcely be for a few hours at sea, or by the water in a harbour-town, without some of them, from curiosity or carelessness, coming round so close to one as to afford proof of this. I was once fishing outside the harbour at Stornaway, and I threw over, with a short rod and line, two gulls which were flying close to us, attracted by a hope of a share in our fish. The first, when released, continued to hover round us, not exhibiting the slightest fear. I took him in my lap and offered him some nice bits of fish. At first he pretended to be angry, and pecked at my fingers, as if to ask whether I thought it possible he would accept my donations under restraint. But having got hold of a piece of fish, down it went; and then, apparently thinking that he might do worse, he set to work in good earnest, and soon got through the most part of a haddock. On regaining his liberty, he continued to fly backwards and forwards, within a few feet of our heads. The captain of a Dover steamer told me he had seen a gull come and take off the taffrail food which had been placed there for him. They are, in fact, the most fearless of birds."

Mr. Drosier, of Norfolk, gives the following description of a fight between some gulls and an eagle :—

"As I was intently observing the majestic flight of the eagle, on a sudden he altered his direction, and descended rapidly, as if in the act of pouncing. In a moment five or six of the gulls passed over my head with astonishing rapidity ; they soon came up with the eagle, and a desperate engagement ensued. The short

bark of the eagle was heard above the cry of the gulls.
They never ventured to attack him in front, but taking
circular flights around him, one of them would make a
desperate sweep or stoop, and striking the eagle on the
back, would dart up again almost perpendicularly,
waiting for an opportunity to repeat the attack. Thus
passing in quick succession, the gulls harassed the
eagle most unmercifully. If, however, he **turned to-**
wards one of them, the gull quickly fled without touching him. This engagement continued for some time,—
the eagle turning and wheeling as well as his ponderous
wings would allow, till he approached some rocks; and
then the gulls made off."

We are not in the habit of attributing such feelings
as **those of** sympathy **and** attachment to such a creature
as a hawk. Mr. Knox, however, **in his book on** " Wild
Fowl," endeavours to correct our ideas in this matter.
He gives us the following story:—

" The late Colonel Johnson, of the Rifle Brigade, was
ordered to Canada with his battalion; and being fond of
falconry, he **took with** him two of his favourite pere-
grine-falcons on **his** voyage. He accustomed them to
a flight every day,—feeding them before starting, that
they might not be led away by the calls of hunger.
Sometimes their rambles were very wide; at others,
theywould ascend to a height which almost defied the

eye to follow them. Still, as they retained the habit of returning to the ship each evening, no anxiety was felt as to their movements. At last, however, after a longer flight than usual, one of the falcons returned alone. The other, the favourite, was missing. Day after day passed, and Colonel Johnson began to make up his mind to the loss; when, arriving in Canada, and casting his eye over a Halifax newspaper, he was struck by a casual mention that the captain of an American schooner, just arrived, had brought with him a fine hawk, which had alighted upon his vessel during his voyage from Liverpool. It occurred to Colonel Johnson that this probably was his much-prized falcon; and hence, asking for leave of absence, he set out for Halifax. On arriving there, he soon found out the American captain, and explained the object of his visit. But the present possessor of the bird was in no hurry to part with it, and boldly declared his disbelief of the whole story. In this difficulty Colonel Johnson could only appeal to the Americans present, asking for a fair trial, in which the bird should show, or should refuse to show, any signs of recognition of her old master. So challenged, the American captain could hardly, with a fair character, evade the trial. He went upstairs, and returned with the falcon. The door was scarcely opened before the bird darted to her old protector, showing in every way her delight, rubbing her head against his cheek, and biting the buttons of his coat in playfulness. The verdict of the bystanders was instantly given, and the bird was restored to its owner."

Bishop Stanley gives us one or two stories of wild and domestic birds, from which we copy these :—

" An officer, settled on a farm near the Missouri, observed one day, when walking near the banks of the river, a large eagle, which seemed to be frequently darting down to the water, and then rising again. Drawing nearer, he perceived that the object of attack was a wild goose, which was floating on the water, and which dived beneath whenever its enemy stooped. But the goose was exhausted, after a time, by these efforts to escape ; and at last it fled to the shore, where two men were at work, to whom it surrendered rather than fall into the talons of the eagle. It found the protection it sought, and in two or three days seemed quite reconciled to its new position."

" A game cock at Ashford, in Kent, priding himself on his valour and prowess, happened to take offence at a goose, who was then sitting, attacked her with great fury, pecked out one of her eyes, and broke several of her eggs. The gander, seeing the danger, came to his mate's assistance, and a battle took place.

The next day the cock renewed his attack upon the goose; when the gander, hearing the quarrel, hurried up, and seizing the cock, dragged him to the pond, where he held him under water till he was actually drowned."

" At Astbury, near Congleton, some geese were feeding near a barn where some men were threshing. A sparrow was near them, pecking up what it could. A hawk saw it, pounced down, and the poor sparrow's end would have been speedy, had not a gander rushed to the spot, and with its beak struck the hawk such a blow that it was stunned, and almost killed.''

" A goose belonging to a clergyman in Cheshire began to sit upon six or eight eggs. The dairymaid, thinking these too few, added some ducks' eggs. The next morning she went to see if all was right, but found all the ducks' eggs picked out and lying on the ground. She replaced them, but next morning the same scene met her view: all the ducks' eggs had been turned out. For fear of disgusting and driving the goose away, the experiment was tried no further."

Mr. Kirkham writes to the *Naturalist*,—
" A friend of mine who, a few years ago, resided at Bank House, Burnley, was in the habit of feeding daily a well-known and favourite robin. An envious sparrow,

coming one day to join in the good things provided, was driven away by the robin ; and, not well pleased at the treatment it had received, came again, attended by superior numbers, so that now it was the red-breast's turn to be defeated. But Robin had a resource. The next day, to the surprise of all, a crow appeared with his red-breasted friend, and kept at a distance all rivals and intruders. The two then partook amicably together of whatever was provided. This continued for several days, and then, something having occurred, they came no more."

"The swan," says Yarrell, in his "British Birds," " exhibited, eight or nine years ago, one of the most remarkable instances of what we call 'instinct' that ever was recorded. She was sitting on four or five eggs, and was observed to be very busy in collecting weeds, grasses, leaves, or what not, to raise her nest. A farm-ing-man was ordered to take down half a load of haulm, with which she most industriously raised her nest, some two feet and a half higher than usual. The next night there came down a tremendous fall of rain, which flooded all the lowlands, and did great damage. Men had made no preparation; the bird had. Her eggs were above, and only just above, the water.

A writer in the " Encyclopædia Britannica " says,—
" Last spring a pair of crows made their nest on a

tree, of which there are several round this garden.   One
result was, a constant warfare between them and a cat.
One morning the battle raged fiercely, till at last the
cat gave way, and took shelter under a hedge, to look
out for some opportunity.   The crows went on for some
time with threatening noises, but they soon saw that
noises went for nothing.   Then, after consideration, one
of the crows took up a stone from the middle of the
garden, and waited with it in her claws till the enemy
should appear.   When the cat began to creep along the
hedge, the crow accompanied her; and when at last puss
ventured forth, the crow, hovering over her in the air,
let fall on her the stone which she held in her claws.
That there must have been some ratiocination in all this
is hardly to be denied."

" A pair of larks," says Mr. Wood, "had built their
nest in a grass field, where they hatched a brood of
young.   Very soon after the young birds were out of
the eggs the owner of the field was forced to set his
mowers to work—the state of the weather obliging him
to cut his grass sooner than usual.   When the labourers
approached the nest, the parent-birds took alarm, and at
last the mother-bird laid herself flat on the ground, with
outspread wings and tail, while the male bird took one
of the young out of the nest, and, by dint of pushing
and pulling, got it on to its mother's back.   She then

flew away with it over the fields, and soon returned for another. This time the father took his turn, and carried one of the young ones, the mother helping to get it on to his back ; and so they managed to remove the whole brood before the mowers had reached their nest."

Mr. Webber describes a ruby-throated humming-bird, which he was fortunate enough to capture alive :—

" It occurred to me that a mixture of two parts of loaf-sugar, with one of honey, and ten of water, would make about the nearest approach to the nectar of flowers. While my sister was preparing it, I gradually opened my hand to look at my prisoner, and found, to my no small amusement, that it was actually 'playing 'possum,' —that is, feigning to be dead. It lay on my open palm motionless for some minutes, during which I watched it with breathless curiosity. I saw it gradually open its bright little eyes, to peep whether the way was clear, and then close them again, when it caught my eye fixed upon it. But when the 'nectar' was ready, and a drop was put upon the point of its bill, it came to life very suddenly, and in a moment was on its legs, drinking eagerly from a silver teaspoon. When it was satisfied, it refused to take any more, and sat with entire self-composure on my finger, pluming and trimming itself as if it had been on its favourite spray. I was enchanted with the bold, innocent confidence with which it turned

up its keen black eyes to survey us, as if it would say,
'Well, good folks, who are you?' Thus in less than
an hour the tameless rider of the winds was perched
pleasantly on my finger, and received its food with
eagerness from my sister's hand. By the next day it
would come from any part of either room, alight on the
side of a white china cup containing the mixture, and
drink eagerly, with its long bill thrust down to the very
bottom. It would alight on our fingers, and seem to
talk with us endearingly in its soft chirps. I never saw
any creature so thoroughly tamed in so short a time.
But after keeping it for about three weeks, it began to
droop, so that I felt obliged to let it have its liberty.
As soon as the window was opened, the bird darted
out like a meteor, and was instantly gone. In hope of
attracting him back again, a fresh cup of 'nectar' was
prepared, and the cage and the cup, hung with flowers,
was placed on the window-sill. After waiting about an
hour, the beautiful creature suddenly appeared, hovering
over the window. He was darting to and fro before
the window, as if the flowers perplexed him. At last
the well-known cup seemed to overcome his doubts,
and he settled upon it, as of old. We rushed forward,
while he was drinking, to secure him; but a better
thought seemed to rebuke our want of trust, and we
threw open the cage-door again, and let him have the
day to himself."

Wilson tells us that without the sun these little creatures will die, or will simulate death :—

" A beautiful male bird was brought to me this season, which I put into a wire cage, and placed in a shaded part of the room. After fluttering about for some time, as if desirous to get out, the weather being cold, it clung to the wires, and hung there in a torpid state for the whole forenoon. I could perceive no motion of the lungs; the eyes were shut, and when touched by the finger, it gave no sign of life or motion. I carried it out into the air, and placed it in the rays of the sun, but sheltered from the wind. In a few moments its respiration became apparent, it breathed faster and faster, opened its eyes, and began to look about. When it had recovered, I gave it liberty, and it flew off to the top of a pear-tree, where it sat for some time pluming itself, and then shot off like a meteor."

Mr. Simeon, in his "Stray Notes," says,—

" I was walking one day with a gentleman on his home farm, when we observed the grass on about an acre of meadow-land to be so completely rooted up and scarified that he took it for granted it had been done under the bailiff's direction, in order to clear it from moss; and when he met the man, he asked whether this was the case. The bailiff replied, ' Oh no, sir; we have not been at work there at all; the rooks have done

all that.' I had often seen places where the grass had
been pulled up by rooks; but I never had seen such
clean and wholesale work as this. I found that the
object of the rooks had been the devouring of a small
white grub, which had settled there in great numbers,
and which had hidden themselves a little below the
surface.''

Dr. Stevelly, of Belfast, thus describes a magpie of
his acquaintance :—

"He was particularly fond of any shining article,
such as spoons and trinkets : these he frequently stole;
and we came upon his hiding-place in a remarkable way.
There was an old gentleman, a friend of my father's,
who resided with us a great deal. He was of a
studious disposition, but was fond of reading, standing
with his arms and breast resting on the back of a chair,
and the book lying on the table before him. After
reading for awhile, he would take off his spectacles, lay
them beside him, blow his nose, take a pinch of snuff,
and then remain pondering what he had read; till, after
a while, he would resume his spectacles and proceed.
One warm day I lay reading at the end of a room out of
which a glass door led to the greenhouse. At a little
distance from me the old gentleman was pursuing his
studies. After a while the magpie perched on a chair
near him, eying him most intently; and at last he

reached, with an active hop, the table, seized the spec-
tacle-case, and was out of the glass door in an instant.
I remained quiet, meaning to see the end of this.
After a short absence, ' Jack ' was at his post again,
eying the old gentleman with a most business-like glance.
Presently off came the spectacles, and out came the
pocket-handkerchief and snuff-box. In a moment ' Jack '
had lighted on the table, and was out by the open door
with his prize. And now he did not return : having
gained the object of his ambition, he thought it wiser to
get wholly out of sight. At length, the old gentleman having concluded his reflections, he wished to
resume his spectacles. As soon as his surprise at not
finding them on the table had abated, he moved his
chair, and searched all around, to find out where they
could be. Of course, when things had reached this
point, I burst into a fit of laughter, and the old gentleman angrily accused me of having played off some
joke. I assured him that I had never stirred from the
sofa ; and after a while I was obliged to explain to
him what I had seen of ' Jack's ' manœuvres. Then the
question arose, how the spectacles were to be recovered.
Somebody proposed to leave a teaspoon in his way.
This was done ; but ' Master Jack ' was too wary,—he
would not approach us. At last, by dint of a most
resolute watch, it was found that he had a place of
deposit in a corner of the roof ; and here not only the

spectacles were found, but sundry other **things which** had for some time been missing."

In the *Northampton Mercury*, of August 1851, **the** following curious anecdote was given :—

" A stoat was making its way **from** an adjoining field, across the road, with a young partridge in its mouth which it had killed, when **it** was attacked and pursued by two skylarks and a wagtail. The three birds, acting together, rose a little in the air, and then pounced down again and again upon their enemy, repeating their attacks so furiously that he was obliged to abandon his prey. He strove to regain it, but their attacks were renewed with the greatest courage ; till at last some one approached, and the stoat ran off, leaving the partridge behind."

The late Mrs. O'Brien, **of Chelsea, had a** canary which was a great favourite, **but** whose loud singing often **obliged** her to put him outside the window among the **trees** in front of the house. One morning, while thus placed, a sparrow took pity on the prisoner. He was observed to fly round and round it, to stand upon the top, and twitter to the bird within, between whom and itself some conversation seemed to pass. After a while the sparrow flew away, but soon returned with a worm in his bill, which he dropped into the cage, and which the canary gladly received. Day after day these visits were repeated, and fresh presents brought; till the canary at last received the food out of the sparrow's bill. The sparrow was shy to human beings, but fond of his own kind. When winter came, the canary was taken in-doors, and the sparrow disappeared.

Dr. Edmonstone tells us that in northern Scotland extraordinary meetings of crows are occasionally seen to occur. They collect in great numbers, as if they had been summoned for the occasion. A few of them sit with drooping heads, and others sit like judges, while others are exceedingly active and noisy. In the course of an hour or two they disperse; but it is not uncommon to find, when they are gone, that one or two have been left dead on the place of meeting. "Crows seem to come from all quarters. When they have all assembled, a general noise or conversation follows, and shortly after the whole fall upon one or two of the number, and put them to death. When this has been done they quietly disperse."

Captain Brown writes:—

"Mr. William Wright, a publican at Gilmerton, near Edinburgh, had a tame jackdaw. On one occasion half a glass of whisky was left on the table, when 'Jackie' flew up, and, after the first taste, liked it so much that he drank greedily. Soon symptoms of intoxication began to appear: his wings dropped, and his eyes were half closed. He staggered in his walk in a most ludicrous manner. He moved to the edge of the table, seeming not to know what to do; till at last his eyes closed, and he fell on his back, showing every sign of death. Water was given him, but he could not swallow it. He was rolled up in flannel, and put in a box. The family never expected to see him again. But next morning, when the door was opened, he flew out, and made his way to a stone basin out of which the fowls drank, where he copiously allayed his thirst. He repeated his draughts of water during the day, but never afterwards would he touch whisky."

Dr. Stanley is obliged to conclude that young cuckoos have some quality which enables them to gain the affections of other birds. As an instance of this, he tells us that—

"A young cuckoo was put into a cage, and, a few days after, a scarcely-fledged thrush was also put in. The thrush could feed itself, but the cuckoo was obliged to be fed with a quill. In a short time, however, the thrush began to feed its fellow-prisoner, bestowing on it every possible attention, and manifesting the greatest anxiety to satisfy its continual craving for food."

He adds another instance :—

" A young thrush, just able to feed itself, was placed in a cage. A short time after, a young cuckoo, which could not **feed** itself, was placed in the same cage, and fed **by the** owner. At length it was observed that the thrush **fed** it—the cuckoo opening its mouth, sitting upon the **upper** perch, and making the thrush hop down to **fetch its** food. One day, while it was thus expecting its breakfast, a worm was put in the cage, and the thrush could not resist the temptation of eating it. Imme-diately the cuckoo descended, attacked the thrush **with** fury, and actually tore out one of its eyes, and then hopped back. Yet even after this the thrush meekly took it **some** food, and continued to do so till the cuckoo was full **grown.**"

Captain Brown tells us of a pigeon **which** belonged to an innkeeper at Cheltenham :—

" He was twelve years old when his partner deserted **him.** He seemed deeply grieved, but made no new **alliance. For two** years he remained widowed and for-saken ; when at last his faithless partner returned, and wished to be taken back again. He repelled her with disgust, and when she became importunate he pecked her severely, and drove her off. But in the course of **the** night she managed to effect a lodgment ; and the next morning he seemed to be so far reconciled as to

allow her a place in his home. But soon afterwards she
died. He now seemed sensible that by her death he
was set at liberty; for he quickly took wing, and in a
few hours returned with a new partner."

The same writer tells us the following story of a
stork :—

"A tame bird of this sort had taken up his abode for
some years in the yard of the college at Tübingen. On
a neighbouring house was a nest, where the storks that
resorted to the place used to hatch their eggs. One
day in autumn a young collegian fired a shot at this
nest. Probably he wounded the stork who was then
sitting. Still, at the usual time, all the storks took their
departure. Next spring a stork appeared on the roof
of the college, and by clapping his wings seemed to
invite the tame stork to come to him. But the latter
was unable, as his wings were clipped. After a time
the wild stork came down into the yard, and the tame
one went to meet him, clapping his wings in welcome ;
but the wild one attacked him with the greatest fury.
The people interfered and drove him away ; but all that
summer he came down into the yard, again and again,
to attack the other. The next spring, instead of one
stork, four of them came down into the yard, and attacked
the tame one : seeing his danger, the cocks, and ganders,
and turkeys rushed to help him, and managed to drive

the assailants away. Greater watch was now kept ; but when the third spring came, twenty storks or more descended into the yard at once, and killed the poor bird before any help could come to him."

Mr. Brew, of Ennis, says,—

"An old goose that had been sitting on her eggs for a fortnight, in a farmer's kitchen, was perceived on a sudden to be taken violently ill. She soon after left the nest, and repaired to an outhouse where there was a goose of the first year, which she brought with her into the kitchen. The young one immediately scrambled into the old one's nest, sat, hatched, and brought up the brood. The old goose, so soon as the young one had taken her place, sat down by the side of the nest, and shortly after died. As the young goose had never been in the habit of entering the kitchen before, I know no other way of accounting for this fact but by supposing that the old bird had some way of communicating her wants, which the other was perfectly able to understand."

Mr. Graves says, in his "British Birds,"—

"A pair of sparrows had built their nest in a wall, close by my house. I noticed that the old birds continued to bring food to the nest, some time after the young brood had left it. I had the curiosity to place a ladder against the wall, and to look into the nest—when, to my surprise, I found a full-grown bird, which had got its leg entangled in some thread, which formed part of the nest, in such a manner as to prevent its getting out. I observed that the parents continued supplying it with food all the autumn, and part of the winter; but when the cold became severe I thought it time to release it. It then flew away with its parents."

VI.

ABROAD.

We have left to the closing chapter of our review some of the most remarkable instances,—without which our investigation would be obviously imperfect. These are, the King of Birds; the most cunning of all creatures, the Monkey; and the Elephant, deemed by many writers the wisest of the brute creation. A few sketches of these remarkable animals will close our retrospect.

The chief disadvantage under which we must labour consists in this, that—especially in the case of the eagle —men have had but few opportunities of placing themselves on friendly terms with these creatures. Yet one case is described by Bishop Stanley, in which a golden eagle, captured while young in Ireland, became quite domesticated, and attached itself to the place in which it dwelt, without being put under any restraint. Its wings had at first been cut, but they were suffered to grow again, and on regaining the full use of them, the noble bird would sometimes soar into the air, and remain absent for two or three weeks. But it became much attached to those who were in the **habit of** feeding or caressing it.   It had been provided with a house, but it generally prepared a perch of its own choosing, on the branch of a large apple-tree.   It fed chiefly upon crows, which were provided for it—being shot for the purpose. It often attempted to catch them; but always in vain, their power of turning rapidly in flight enabling them to escape the eagle's pursuit.   It lived thus in one home for ten or twelve years, and was killed at last in a fight with a mastiff, at which no one was present, but in which the dog was so much injured that he died immediately after.

Dr. Stanley **also tells** us of an extraordinary device practised by **the** eagles in Heligoland, an island of Denmark :—

" Persons resident there state that the eagle will plunge into the sea, so as to wet its feathers, and then roll on the sand till it is quite covered with it.   It will then rise and hover over its victim, generally an ox; shaking itself and scattering the sand into its eyes, till the animal, blinded and terrified, falls into some hollow or over some cliff, and so becomes a prey to its pursuer."

Mr. Thompson writes,—

" My friend, R. Langtrey, Esq., of Fort William,

near Belfast, has a golden eagle, which is extremely
docile and tractable. It was taken from a nest in
Invernesshire, and came into his possession about the
end of September. It soon became attached to its
owner, who, after about a month, ventured to give it
liberty; a privilege of which it took no undue advantage.
It not only permits itself to be handled, but seems to
derive pleasure from its master's caresses. When it is
at large, Mr. S. has only to hold out his arm, and it flies
to him and perches on it. It usually keeps its master
in view, following him round the grounds, and keeping
him in sight."

Audubon, the naturalist, was engaged in collecting
cray-fish, near the Green River in Kentucky, when he
came upon a quantity of white ordure, which his companion, who lived near the spot, said must be from the
nest of the brown eagle, and added that he had seen
one of those birds a few days before. The naturalist
seated himself near the foot of the rock, in high expectation. Two hours elapsed before the parent-bird made
his appearance, which was announced by the hissing
noise of two young ones, who crawled out from the
hidden nest to receive a fine fish. In a few minutes the
mother arrived, bringing also a fish; but her watchful
eye detected the observers; and uttering a loud cry,
she alarmed her partner, and both of them hovered

over the heads of Audubon and his friend, keeping up threatening cries, to scare them from the neighbourhood of the nest. Still, they pressed on, and found the nest; but the young ones had hidden themselves. They watched for a long time in vain; and returning the next day, they found the place deserted.

We return to England, and even to the English metropolis. On the 9th of April, 1848, a magnificent eagle suddenly appeared, sailing over the towers of Westminster Abbey; and, after making various circles, it perched upon the summit of one of the pinnacles. It formed a striking object, and a crowd soon collected to gaze upon it. Presently the bird rose, and began ascending by successive circles to an immense height, and then floated off to the north of London, occasionally giving a gentle flap with his wings, but otherwise appearing to sail away to the clouds, among which he was ultimately lost. The history of this strange appearance was this. Early in that year a white-tailed sea-eagle was brought to London in a Scotch steamer, cooped up in a crib used for wine-bottles, and presenting a most melancholy and forlorn appearance. Mr. F. Buckland, happening to see him, took pity on him, purchased him, and took him to Oxford, where he soon regained his natural noble aspect, delighting especially to dip and wash himself in water, and then to sit on his perch, expanding his magnificent wings, basking in the sun, his head always turned towards that luminary, whose glare he did not fear. A few nights after his arrival at Oxford, we were aroused by cries. The night was dark, but at length the cause of the outcry was discovered. On the grass-plat was the eagle, cowering over a victim with outspread wings, and croaking a defiance to all intruders. Lights being brought, the eagle retired, with his prey, to a dark corner, where he was left to enjoy it. It turned out that he had espied a hedge-hog, and had managed to uncoil him with his sharp bill, and to devour him. The eagle soon became the terror of all the smaller animals in the house. A beautiful little kitten soon disappeared; several guinea-pigs and a jackdaw shared their fate; and " Jacko," the monkey, only saved his life by swiftness of foot. One fine summer's morning the window of the breakfast-room was thrown open, previous to the appearance of the family. Soon the lady of the house appeared, and startling was the sight. A fine ham stood in the middle of the table, and on it was perched the eagle, tearing away at it with boundless appetite, his talons firmly fixed in the rich deep fat. Being disturbed, the eagle's first attempt was to fly off with his prize; but finding it too heavy, he dropped it on the carpet, snatched up a cold partridge, and made a hasty exit through the window. Some time after this the eagle was taken to London, and placed in a court-

yard near Westminster Abbey, where he dwelt in solitary
majesty.   It was from this court-yard that **he** made his
escape on the 9th of April.   He just managed **to** flutter
up to the top of the wall, and from thence he ascended,
with difficulty, until he had cleared the houses ; but as
he ascended into mid-air, his strength returned, and he
soared, in the usual style of an eagle.   His owner said,
" Well, I've seen the last of him ; " but, hoping that he
might possibly find his way back, a chicken was tied to
a stick in the court-yard, and just before dark the
eagle came down, seeking for food, and speedily devoured his prey.   While he was thus engaged, a plaid
was thrown over him, and he was easily secured.   After
this one escape **he** was soon handed over to the Zoological Gardens.

A remarkable story is told by Mr. Blaine, of the
patience, discretion, and self-command of a golden
eagle, **which was** for some time **in the** *Jardin des Plantes*
at Paris.   The bird was taken **in a** trap for foxes, which
broke his claw, and effectually lamed him.   The cure
was a tedious one, and a painful operation was necessary,
which was borne by the eagle with a patience seldom
equalled by human sufferers.   While the operation was
being performed, his head was left at liberty ; but he
offered no opposition to the handling or dressing of the
wounded limb.   Then, swathed in a napkin, and laid on

his side, he passed the whole night without motion, The next day, when the bandages were loosened, he placed himself on a screen, where he remained for twelve hours without once resting on the wounded foot. He made no attempt to escape, though the windows were often open. For several days he remained without food; at last he received eagerly a rabbit which was given him. Having eaten it, he resumed his quiet position, and not until the twenty-first day after his accident did he begin to try the wounded limb. He gradually regained the use of it by cautious and moderate exercise. He passed three months in the room occupied by the man who attended to him. In the morning, as soon as the fire was lighted, he came up to it, and suffered himself to be caressed; in the evening he mounted his screen, where he passed the night. No one could have shown more resignation, more courage,—it might almost be said, more *reason*,— than was shown by this bird during his long illness.

We pass on, from the noble and majestic eagle, to a very dissimilar creature—the grotesque, nimble, and cunning monkey. No two creatures could be more unlike, yet each has its own point of interest.

A friend of Mr. Wood's gave him the following account of a pet monkey, which caused considerable annoyance to those who lived in its neighbourhood:—

" He was one of the **most imitative and** mischievous little animals that ever existed. His imitative tricks caused the servants so much trouble, that he had not a friend among them. One day he observed the ladies'-maid washing her mistress's lace, and his offers of assistance having been somewhat roughly repulsed, he went, chattering and scolding, in search of adventures. Our house was next door, and **as the** windows were open, he **got** in, with the idea of **washing** fresh in his head. Curiosity led him to open **two small drawers,** from which he soon abstracted their **whole contents—** lace, ribbons, and handkerchiefs. He threw all these things into a foot-pan, pouring in all the water and soap that happened to be in the room; and he must then **have** washed away **with** great vigour, for when I returned to my room, after an hour or two's absence, I found him, to my astonishment, busily engaged in **his** laundry operations, spreading **out** the **torn** and disfigured **remnants to dry.** He **must** have been conscious that he was doing wrong, for, without a word from me, he made off instantly, and hid himself in the clock-case in the kitchen in his own home. The servants saw that he had been in some mischief, for this was his constant place of refuge when he was in disgrace or trouble. One day he watched the cook while she was preparing some partridges for dinner, and apparently came to the conclusion that all birds ought to be so treated; for he

THE CAT'S PAW.

managed to get into the yard, where his mistress kept a few bantam fowls, and, after robbing them of their eggs, he secured one of the poor hens, with which he proceeded to the kitchen, and at once commenced plucking it. The noise that the poor bird made brought some of the servants to the rescue; but they found the bird in such a pitiable and bleeding state that in mercy it was at once killed. After this the monkey was chained up, which so mortified him that he refused his food, and soon after died."

Dr. Guthrie tells the following anecdote of a monkey:—

"'Jack,' as he was called, seeing his master and some of his friends drinking, with the imitative faculty for which all monkeys are remarkable, got hold of a glass half full of whisky, and drank it off. Of course it flew to his head, and very soon 'Jack' was drunk. Next day, when they wished for a repetition of the performance, he was nowhere to be seen. At last he was found, curled up in a corner of his box. At his master's call he reluctantly came out, but one hand applied to his head signified very plainly that he was ill—that 'Jack' had got a headache. So they left him for a few days, to recover. Then, supposing him to be well again, they called him to join them in another jovial party, expecting to have 'rare fun' with him. But he eyed the glasses with evident dread, and when his master tried to induce him to drink, he was **upon** the house-top **in** a moment.   They called him to come down; but he refused.   His master shook a whip at him; **but** it had no effect.   A gun was then pointed at him, **he** got behind a stack of chimneys.   At length, in fear **of** being dragged from his refuge, he actually descended the chimney; risking a scorching rather than be made to drink.   At last his master gave **way, and** though 'Jack' lived for twelve years after, his repugnance to whisky remained **as strong as ever,** and his master gave way to it."

"My monkey, 'Jacko,'" says **Mr.** Buckland, "was a pretty little fellow; his bright eyes sparkled like two diamonds from beneath his deep-set eyebrows; his teeth were of pearly whiteness, and of these, whether through pride or a wish to intimidate, **he** made a formidable display.   Having arrived at **the** hotel, it became **a question as** to what was to become of 'Jacko' while his **master** was away.   A little closet, opening out of the bedroom, offered itself as a temporary prison. 'Jacko' was tied securely (I thought) to one of the rows **of** pegs that **ran** along the wall.   As the door closed upon him, his eyes seemed to exclaim, 'Now I'll do some mischief!' and sure enough he did.   When I came back to release him, the walls, which half an hour before were covered with an ornamented paper, now stood out in the nakedness of lath and plaster; the pegs were all loosened, and an unfortunate garment that happened to be hung in the closet was torn into a thousand pieces. When 'Jacko' arrived at his ultimate destination, a comfortable home was provided for him in the stall of a stable, where was an aperture communicating with the hay-loft; so that he could either sleep at his ease in the regions above, or, descending into the manger, amuse himself by tearing to pieces everything he could get at. But after a while the donkey was brought out of the field and placed in these comfortable quarters. A supper of hay was placed before the hungry animal, which it began to devour with great eagerness. An hour after, the groom happened to go into the stable to see that all was right. But now the donkey was pulling away at her halter, and trying to keep her head as far as possible from the bundle of hay. Soon the mystery was explained. When the donkey approached the hay a tiny pair of hands were thrust out, and the ears were seized. The little rascal, looking down from the loft, had seen the hay spread out, and had thought that it would make a capital bed for himself, of which he had therefore taken possession, quite regardless of poor Jenny. After a while 'Jacko' got loose. He now quietly sneaked into the knife-house, and tried his hand at cleaning the knives. But the handles were the parts he attempted to polish on the brick-board, and a cut

was found in the middle of his hand next day. He next
set to work to clean the shoes, in imitation of the man
who had charge of him. He covered a shoe **all over,**
sole and all, with blacking, and then he emptied all
that was left in the bottle into the inside of the shoe,
nearly filling it. A day or two after, when the servants
returned from the parlour into the kitchen, they found
that 'Jacko' had taken all the candlesticks out of the
cupboard, and arranged them on the fender, before the
fire, as he had seen done before. Finding the black-lead
in the same place, he took it to a bowl of water, wetted
it, and was diligently rubbing the table all over with it
when he was caught in **the act.** On **the entrance** of the
servants he retreated to his basket in the corner, trying
to look as if nothing had happened."

"I have two monkeys," says the same writer, "of whom
I am exceedingly fond. One is 'The Hag,' and the
other is 'Tiny.' 'The Hag' was originally called
'Fanny;' but she has so much of the disagreeable old
woman about her that I always call her 'The Hag.'
'Tiny' is a very little monkey indeed, not much
bigger than a large rat. She was turned out of the
Zoological Gardens as a dead monkey; she was indeed
'as good as dead,' a mere skeleton, and with very
little hair on her. She arrived tied up in an old canvas
bag. I put her into 'The Hag's' cage. The old lady

at once took to her, and assumed the office of nurse. She cuddled up poor 'Tiny' in her arms, and showed her teeth to anybody who came near. They had port-wine-negus, quinine, beef-tea, egg and milk, and everything she could eat; and 'The Hag' always gave her the first offer of everything. After awhile 'Tiny' could stand; then she could run; her hair came again, and she is now one of the most wicked, intelligent, pretty little beasts that ever committed an act of theft. Steal!—why her whole life is devoted to stealing, for the pure love of the thing. The moment I come down to breakfast I let out the monkeys. I keep a box of sardines especially for 'The Hag,' who immediately helps herself, and sits on the table grunting with pleasure as she licks her oily fingers. The moment 'Tiny' is let loose she steals whatever is on the table; and it is great fun to see her snatch off the red-herring from the plate, and run off with it to the top of the book-shelves. While I am recovering my herring, 'Tiny' runs to the breakfast-table again, and, if she can, she steals the egg. This she tucks under her arm and bolts away, running on her hind-legs. But she is rather fearful of eggs, for she once stole one that was quite hot, and burnt herself. She cried out, and 'The Hag' left off eating sardines, shook her tail violently, and opened her mouth at me, as much as to say, 'You dare hurt my Tiny!' If I keep too sharp a look out upon 'Miss Tiny,' she will run

like a rabbit across the table, and upset whatever she can.
She generally tries the sugar first, as she can then steal
a bit; or she will just pull over the milk-jug. 'Tiny'
and 'The Hag' sometimes go out stealing together.
They climb up my coat, and search all my pockets. I
generally carry some cedar-pencils; the monkeys get
hold of these and bite off the cut ends. But the prime
delight is to pull and try the door of a glass cupboard
till it comes open, and then they can get at the hair-oil,
which they know is there. Any new thing that comes,
they must examine, and when a hamper arrives I let the
monkeys unpack it, especially if I know that it contains
game. They generally end by upsetting the basket,
and then run away crying, 'There's something
alive!' Once I received a snake in a basket, and I let
the monkeys unpack it; I knew that they had a mortal
horror of snakes. When they found out what was
inside, they flew off as fast and as far as possible, crying out, in their language, 'Murder! thieves!' The
parrot talks at them, but they care nothing for old Poll.
'Tiny' steals her seed, and while she runs after 'Tiny'
'The Hag' pulls her tail. 'Tiny' is very attentive to her
old friend. She takes much of what she steals to 'The
Hag's' cage, and is fond of poking paper through the
bars. These things the old monkey tears up, merely to
pass away the time."

Professor Cope had a tame monkey, one of the common Capuchin sort. The creature was kept in a cage, or, rather, was supposed to be kept in it; but it had a strong objection to confinement, and never failed to get loose sooner or later. He generally directed his attention to the hinges; and no matter how firmly they were fixed, the monkey always managed, before long, to extract the staples, pull out the nails, and so open the door at the hinges and not at the latch. Finding that the cage could not hold him, his master had him confined by a strap fastened round his waist, after the manner of monkeys. But the strap proved of no more use than the cage, for the monkey soon contrived to open it. This he did by picking out the threads by which the straps were sewn to the buckles, and so rendering the fastenings useless. But having rid himself of the strap, he thought that he might apply it to some practical purpose. Having perceived that some food had fallen out of his reach, he took the strap, threw it over the morsel of food, and so drew it towards him.

"A great humble-bee," says Mr. Wood, "that had maimed itself, was pushed into the monkeys' cage. Of course it set up a tremendous buzzing, which instantly drew the attention of the monkeys. They were evidently much alarmed at the entrance of such a strange intruder. They approached it with great care, and evident trepidation.   At last, one of them, having well considered the matter, picked up a piece of paper which some one had thrown into the cage, and dextrously twisted it up into a sugar-loaf form.   He then carefully approached the humble-bee, which was lying on its back, spinning round and round, and making a great hubbub, and swept it in an instant into the paper-receptacle, twisting it up without a moment's hesitation, and then patted it and rolled it about till the poor bee's noise was effectually stopped, by its being mashed into a pulp.   And then, gathering courage, the monkey boldly snatched up the paper, and flung it through the bars of the cage."

There are several sorts of animals, possessing cleverness in various degrees, among whom there is perceptible a strong sense of humour.   Thus a French traveller, M. Mouhot, speaks of amusing scenes he has witnessed between the monkey and the crocodile,—

" The latter will be seen lying half-asleep on the bank of a river, and is espied by a crowd of monkeys, who inhabit the trees on the bank.   They seem to consult, to approach, to draw back, and at last to proceed to overt acts of annoyance.   If a convenient branch is within reach, a monkey will go along it, will swing himself down by the end of it, hanging by a hand or a foot, till he can reach to deal the crocodile a slap on the nose, instantly scrambling up the branch, so as to be far out of the enraged brute's reach. Sometimes, if the branch be not near enough, or sufficient, several monkeys will hang to each other, so as to form a chain, and then, swinging backwards and forwards over the crocodile's head, the lowermost monkey will torment the creature to his heart's content. Sometimes the crocodile is so far irritated as to open its enormous jaws, and make a snap at the monkey, just missing him. Then are heard screams and chatterings of exultation among the monkeys, and great gambols are executed among the branches.

Mrs. Bowdich, in the *Magazine of Natural History*, thus describes a monkey of her acquaintance,—

" He was a native of the Gold Coast, and was of the Diana species. He had been purchased by the cook of the vessel in which I sailed from Africa, and was considered his exclusive property. I had embarked with more than a mere womanly aversion to monkeys,—mine was absolute antipathy,—and though I often laughed at ' Jack's ' freaks, still I kept out of his way, till a circumstance brought with it a closer acquaintance, and cured me of my dislike. I was sitting alone on the deck, reading, when, in an instant, something jumped upon my shoulders, twisted its tail round my neck, and screamed close to my ears. My consciousness that it was ' Jack '

scarcely relieved me; but there was no help. I dared
not cry for assistance, because I was afraid of him; and
I dared not try to beat him off, for the same reason. I
therefore became civil from necessity, and from that
moment an alliance between 'Jack' and me commenced.
He gradually loosed his hold, looked into my face, examined my hands and rings with the most minute attention, and soon discovered the biscuit that lay by my side.
When I became reconciled to his society, he became a
constant source of amusement. Like all other nautical
monkeys, he was fond of pulling off the men's caps as
they slept, and throwing them into the sea; of knocking
over the parrots' cages, to drink the water as it ran over
the deck; of stealing the carpenter's tools; in short, of
teasing everything and everybody. Whenever the pigs
were let out, to take a run on deck, he took his station
behind a cask, whence he leaped on the back of one of
them as it passed. The nails he stuck in the pig's back
to keep himself on, produced a squealing, but 'Jack' was
never thrown. Confinement was the worst punishment
he could receive, and when threatened with it, he would
cling to me for protection. At night, when about to
be sent to bed in an empty hen-coop, he generally hid
himself under my shawl, and at last no one but I could
put him to rest. He was particularly jealous of the
other monkeys on board, and managed to put two of
them out of the way. Of course he was scolded and

flogged. But his spite against his own race was manifested at another time in a very original way. The men had been painting the ship's side with a streak of white, and being called to dinner, left their brushes and paint on deck. He called a little black monkey to him, who, like the others, immediately crouched to his superior. 'Jack' seized him by the nape of the neck, took the brush, dripping with paint, and covered him with white from head to foot. Both the man at the helm and I burst into a laugh; upon which 'Jack' dropped his victim, and scampered up the rigging. The unhappy little beast began licking himself, but I called the steward, who washed him so well with turpentine that all injury was prevented. 'Jack' was peeping through the bars of the maintop, apparently enjoying the confusion. For three days he remained aloft, fearful of a flogging. At last hunger brought him down. He dropped, from some height, upon my knees, and I could not deliver him up to punishment."

Mrs. Lee relates that, having annoyed one of the monkeys in the *Jardin des Plantes*, in Paris, by preventing him from taking the food of one of his companions, and having given him a knock on the paw, the creature never forgave the offence. Whenever she came into the garden, at subsequent times, he would show the greatest rage so soon as he heard her voice. And when,

14

after the lapse of months, she again visited the place, happening to approach too near to his cage, he seized upon her gown, dragged the skirt of it within his bars, and bit a large piece out of it.

Another monkey showed the same gift of memory, but in a more pleasing way. Mrs. Lee had brought him from Senegambia, and not finding him a pleasant inmate in a European house, she gave him to the *Jardin des Plantes.* While she stayed in Paris she paid him frequent visits. When she left, the monkey showed, as the keeper afterwards told her, great disappointment, constantly watching for her return. Two years afterwards she revisited the place, and sought him out, and when she said, " ' Mac,' do you know me? " he gave a scream of delight, put both his paws through the bars, stretching them out to her, held his head down to be caressed, uttering a low murmur of pleasure, and showing every sign of delight at the meeting.

In Mr. Forbes's " Oriental Memoirs " we read,—

" On a shooting party one of my friends killed a female monkey, and carried it to his tent, which was soon surrounded by forty or fifty of the tribe, who made a great noise, and advanced in a menacing manner. When he presented his fowling-piece, they retreated; but one stood his ground, chattering and menacing in a furious manner. At length he came nearer to the tent-

door, and finding that his threatenings were of no avail,
he began a lamentable moaning, and by every expression of grief seemed to beg for the body. It was given
to him. He took it up in his arms, eagerly pressed it to
him, and carried it off in a sort of triumph to his companions."

An orang-outang, or Great Monkey of Asia, was
procured by Captain Methven from the south coast of
Borneo. He was put on board the *Cæsar*, and while left
at liberty made no attempt to escape, but when put
into confinement, became violent, shaking the rails of
his cage, in an effort to break them. Finding that they
did not yield, he tried them one by one, and perceiving
one to be weaker than the rest, he worked at it unceasingly till he had broken it, and then he forced his
way out. After several attempts to secure him effectually, he was at last allowed to wander freely about the
ship, and then he soon became familiar with all the
sailors. They often chased him, for fun, about the
rigging; but in dexterity he soon shewed himself their
superior. Sometimes he would outstrip his pursuers by
mere speed; and when pressed, he would catch hold of
a loose rope, and swing himself out of their reach. At
other times he would wait patiently till his pursuers
almost reached him, and then suddenly seize a rope
which hung near him, and lower himself to the deck

while they were still on the shrouds or at the mast-head.
Sometimes, in a playful humour, swinging by a rope, he
would swing within arm's length of his pursuer, and
then, dealing him a blow on the face with his hand,
would fly rapidly away. While on board ship, he usually
slept at the mast-head, wrapping himself in a sail. If a
sail was not to be got, he would run off with one of the
sailor's jackets. In cold latitudes he suffered much
from the weather, coming down from the mast in the
morning shivering with cold, and glad to climb into the
arms of any of his friends, to get warmth from them.
He was generally eager for food, and would follow an
acquaintance all over the ship to obtain it. He would
eagerly rifle and empty the pockets. He soon became
strongly attached to those who used him kindly; sitting,
whenever he could, by their side, and taking their hands
to his lips. From the boatswain he learnt to eat with a
spoon, and might often be seen at his cabin-door,
enjoying his coffee, and seeming quite indifferent to
the gaze or the remarks of the bystanders.

"Next to the boatswain," says Dr. Abel, " I was
perhaps his most intimate acquaintance. He would
often follow me to the mast-head, when I went there for
an hour's quiet reading; and having satisfied himself
that my pockets contained no eatables, he would lie
down, and, pulling a topsail over him, would quietly

wait and watch my movements. On shore his favourite amusement was swinging from the branches of trees, passing from one tree to another, and in climbing over the roofs of houses. While on board ship, he delighted in romping with the ship-boys. To provoke them to a game, he would strike them with his hand in passing, and then, springing away, would engage in a race and a scuffle, in which he used his hands, his feet, and his mouth. Three small monkeys from Java were brought on board; but of them he took little notice, not treating them as equals, nor condescending to romp with them, as he did with the boys of the ship. The monkeys, however, had evidently a great desire for his company, for whenever they got loose they always found their way to his resting-place, and tried to gain his protection. Once only he showed violent alarm. At the Island of Ascension eight large turtles were brought on board, and so soon as he caught sight of them he climbed with all possible speed to the highest point of the mast that he could reach, and looking down upon the turtles, uttered sounds like the grunting of a pig. After a while he ventured cautiously to descend, but could not be induced to come within many yards of them."

A red orang-outang exhibited in Edinburgh gave a proof of intelligence and dexterity. His master gave him one day the half of an orange, a fruit of which he was very fond, putting the other half on an upper shelf, out of his reach and sight, a proceeding of which the creature seemed to take no notice. But when his master threw himself on the sofa, and closed his eyes, the orang began to prowl about and to plan. He first came near the sofa, and tried to be sure that his master was really asleep. The man, who was watching him, succeeded in preserving the appearance of slumber. The monkey then ventured to climb up to the shelf, and getting possession of the rest of the orange, quickly devoured it, all but the peel, which he carefully concealed in the grate, among some paper and shavings.

An Indian paper tells the following story,—

" A Madrassee had a monkey which he was very fond of. The man had occasion to go on a journey, and took with him money and jewels, and his chum, the monkey. Some rogues determined to rob him of everything he had ; accordingly they lay in wait for him and murdered him. Having secured the money and jewels, they threw the murdered man into a dry well, and having covered the body with twigs and dry leaves, they went home. The monkey, who was on the top of a tree, saw the whole proceedings, and when the murderers departed, he came down, and made his way to the tahsildar's house, and by his sighs and moans attracted the attention of that functionary. Inviting the tahsildar

by dumb signs to follow him, the monkey went to the
well and pointed downwards. The tahsildar thereupon
got men to go down, and, of course, the body was discovered. The monkey then led the men to the place
where the jewels and money were buried. He next
took them to the bazaars, and as he caught sight of one
of the murderers he ran after him, bit him in the leg,
and would not let him go till he was secured. In this
way all the murderers were caught. The men, it is
said, have confessed their crime, and they now stand
committed for trial before the Tellicherry Court, at the
ensuing session. That monkey ought to be made an
inspector of police."

Many anecdotes have been published of the memory
of the Elephant, and of its exhibition of anger, or of
gratitude. The variety and frequency of these anecdotes tend to confirm their truth. That which has
been observed only at one time, and in one place, may
be doubted ; but if similar relations are offered, from
different places, by different persons, the various stories
give a probability, and therefore a confirmation, to each
other.

Thus, the narrative of the tailor at Delhi has been
known for more than half a century. He had been in
the habit of gratifying an elephant that daily passed by
his shop, with an apple, or some other acceptable eat-

able, till the animal had grown quite accustomed to the gift. But one day, being out of humour, **the** tailor, when the elephant thrust in its trunk to **receive** the customary gift, thrust his needle into the proboscis, as a warning to be gone. The creature passed on, disappointed, but not forgetting. On its return, passing a pool of dirty water, it filled its trunk, and on coming to the tailor's shop-window, discharged the muddy water over the poor artisan.

Just such another anecdote is told of another elephant, who was kept in the menagerie at Versailles. A painter, **desirous of** drawing the creature with its trunk uplifted, **desired** a boy **to** encourage the creature to take this posture by throwing, or offering to throw, fruit into his mouth. But being deceived more than once or twice, it grew angry, and, with elephantine understanding, it did not punish the boy, but the painter who employed him. It filled its trunk with water, and discharged it over the paper **or canvas on which** the painter was trying to delineate it.

In another case, in the present menagerie at Paris, a sentinel was very careful to warn the spectators not to give the elephant anything to eat. This caused great displeasure in the animal's mind, who soon understood that it was the sentinel that kept food from being given.

One day, filling her trunk with water, she showed her resentment, so soon as he interposed, by discharging over him a shower of water. The spectators laughed loudly, but the sentinel was not to be daunted. He presently renewed his cautions to fresh visitors; but now the elephant, growing still more angry, seized his musket with her trunk, trod it under her feet, and did not allow him to recover it till she had made it wholly useless.

Mr. Jesse tells us that he was one day feeding the elephant at Exeter 'Change (now long since demolished) with potatoes, which the animal took eagerly out of his hand. One of them, at last, fell on the floor, just out of the reach of the creature's proboscis. After several fruitless attempts to reach it, he seemed to give a few moments to consideration, and then blew so violently against the potato that it was driven to the opposite wall, and thence rebounding, came within reach of the animal's trunk.

Buffon quotes from M. Phillipe, who was himself an eye-witness, a description of the workmanlike qualities of an elephant at Goa. A large ship was building in that place, and the elephant was employed to carry down the timber. A heavy beam was bound with a rope, and the rope was then handed to the elephant. Often it was so

large that twenty men could scarcely have moved it.   But
alone, without any direction, the elephant dragged it
down to the place where he knew that it was wanted.
Some obstructions would present themselves, and at
once, like a human being, the elephant would lift the end
of his beam over others, or edge it forward, precisely as
an experienced carpenter or builder would have done.

The elephant will also imitate, or from his own consciousness adopt, a similar course to that taken by men, in
appealing to his strength and power.   Thus, some young
camels, travelling with the British army in India, came to
the river Jumna, and a boat was provided to carry them
across.   But they were frightened, and refused to go into
the boat.   An elephant, who was in the expedition, was
brought forward, and seemed at once to comprehend the
difficulty.   He instantly took a posture of command,
made a loud trumpeting, shook his ears, roared, struck
the ground vehemently, and made the dust rise in clouds.
The poor camels, in a fright, were but too glad to rush
into the boat to get out of his way.   And then at once
the elephant resumed his wonted tranquillity, and all was
over.

In one of the Indian campaigns in Scinde and the
Punjaub, a body of artillery was proceeding up a hilly
road, and the elephants were of great use in drawing

the heavy guns. On one of the gun-carriages sat an artilleryman, resting himself, in the hot and fatiguing day. Just behind him followed another gun-carriage, drawn also by an elephant. The artilleryman on the first, overcome by the heat, slipped from his seat, and fell upon the road. The second, following, would, apparently, pass over and crush him. But the elephant of the second, seeing the danger, and being unable to reach the man, seized the wheel of the carriage he was drawing, and lifted it up, so as to pass over the poor man on the road, and then dropped it on the other side.

An officer, out upon a field-sport excursion in India, was traversing a jungle, on the back of a favourite elephant. All at once he saw a tiger, crouching for a spring, just a little in advance of the elephant's head. Instantly he shouldered his rifle; it missed. He threw it on the ground to seize another, when, to his surprise, the elephant picked up the fallen gun with his trunk, and returned it to him, as if aware that it might soon be wanted.

On another occasion, the same gentleman, when tiger-hunting, perceived that a tiger was lurking in a thick jungle near to him, but from which it was not easy to expel him. The mahout, or driver, was bidden to let the elephant know that he was to beat the bushes with a branch of a tree. This he did so vigorously that the tiger sprang out, and was shot. The sportsman was so pleased, that he told the mahout to give the elephant some sugar (of which those creatures are very fond) when they got home. The mahout forgot this order ; but in the evening the elephant found out his master, rubbed him with his trunk, and gave him evident signs that the promise was not forgotten.

On one occasion an elephant was ordered to drag a tree from the place where it lay; but the weight proved too great for its strength. At last the chains or cords broke, and the elephant was free. It fled, and it was supposed that it had gone into the jungle, and to mix with the wild elephants who wandered there. But instead of this, it returned, in an hour or two, and with it came two other elephants; and these, by their united strength, proved able to perform the task.

Mr. Swainson gives an anecdote of an elephant who always refused to carry what he deemed too heavy a burden. If more than enough was piled up, he pulled it off with his trunk. One day a quarter-master, irritated at what he deemed perverseness, threw a tent-pole at his head, and walked away. The elephant did not then pursue him, but meeting him a few days after, on the public road, he lifted him up into a large tamarind-tree, and left him to get down how he could.

Another elephant, while passing through London, was followed by a man who took hold of his tail. He resented the indignity by turning round, seizing the man with his trunk, and pressing him up against some iron rails, from which he was only released by the keeper's interference.

The *Bristol Times* avouches the truth of the following narrative :—

" A few days ago a fine Asiatic elephant was landed at Southampton, and purchased by Mr. A. Fairgrieve. It had been shipped at Bombay, and had been placed under the care of a Sumatran named Rainee Jhandegar, to whom the creature was particularly attached, and who had brought it into perfect subjection. During the voyage, for the first few days everything went on satisfactorily; but when bad weather came, a spirit of insubordination appeared among the crew. Some of them, contemplating mutiny, used to meet near the wheel-house on the main deck, close to the elephant and to Rainee. There they talked over their plans. The Sumatran feigned to be asleep; but he listened to what the mutineers were saying. He heard them plan the murder of the captain and the chief of the crew;

and he heard that the attempt was to be made that
night. It seemed necessary to warn the captain; but
how to do it he knew not, for it would have been
necessary to pass through the mutineers, who would not
have hesitated to murder any one who seemed about to
give an alarm. He took the only course possible. He
set the elephant at liberty. Springing into the midst of
the mutineers, and giving a sign to the elephant, the
creature laid about it with its trunk, and the surprised
sailors were quickly prostrated on the deck, wounded,
and calling for help, and for mercy. The captain, hearing the noise, was soon in the midst of them, and Rainee
explained what had occurred. The mutineers were at
once placed in irons, and order was restored. During
the same voyage the elephant performed other services.
Once, during a heavy gale, he saw a man slipping off
the bulwarks, and saved him from a watery grave by
seizing hold of his jacket. On another occasion, an
angry mastiff was flying at the throat of the mate; when
the elephant, taking hold of him with his trunk, hurled
him over the side of the ship into the sea.

In 1869, at Mongghyr, in India, a railway-train approached a mango-tope, in which about seventy elephants
were stationed. This unusual sight, the red lights, the
smoke, and the noise, produced a terrible commotion
among the poor brutes, who all attempted to break

away from their fastenings. One of the largest and strongest succeeded in breaking loose, and he rushed forward to oppose the approaching enemy. He mounted the line just as the train came up—he charged it with head and tusks; but the power of steam and machinery was too great. He was knocked down and killed, but the train was thrown off the line, and eleven carriages were capsized into a ditch. The injuries to the passengers were not many.

Sir Emerson Tennant mentions an incident which occurred to him in Ceylon.

" One evening, while riding in the vicinity of Kandy, through a jungle, my horse showed some alarm at a sound which approached us, and which consisted in a cry of ' *Urmph ! urmph !*' in a hoarse and dissatisfied tone. Suddenly we came upon a tame elephant, who was labouring painfully to carry a heavy beam of timber, which he balanced across his tusks ; but the pathway being narrow, he was forced to bend his head to allow it to pass end-ways, and the trouble this gave him called forth the noise and grumbling which had alarmed us. On seeing us halt, the elephant raised his head, looked at us, and then flung down the timber, and forced himself backward among the brushwood so as to leave a passage, of which he expected us to avail ourselves. My horse still hesitated, and the elephant, observing

this, thrust himself still deeper into the jungle, en-
couraging us to come on. Still the horse trembled,
and again the elephant wedged himself still farther in
among the trees, and waited impatiently for us to pass
him. When the horse had at last done so, I looked
back, and saw the sagacious creature stoop and take
up again his heavy burden, balancing it on his tusks,
and resuming his way, snorting, as before, his discontented remonstrance."

In Pondicherry, in the East Indies, an elephant,
whose keeper was for a short time absent, was amusing
himself while waiting, when a man who had committed
a theft, and was pursued by the officers, and by a crowd
of people, threw himself upon the protection of this
powerful animal, creeping under his belly. The elephant seemed pleased with this mark of confidence, and
seemed at once to take the fugitive under his care. He
raised his trunk in a threatening manner, and showed
himself so determined in the fugitive's defence that
neither the crowd, nor even his keeper, who entered
into the strife, could prevail upon him to give up the
man. The contest continued for a long time, till at
last the governor, hearing of it, came to the spot, and
was so much struck by the generous zeal of the animal
that he pardoned the criminal, and, with the keeper's
help, made the elephant understand that the man was

safe. The pardoned offender, in an ecstacy of gratitude, embraced the trunk of the elephant, who apparently was so well satisfied that he became perfectly
quiet, and went away with his keeper as placidly as if
nothing had happened.

The Baron de Lauriston relates that when an epi-
demic disorder was raging at Lucknow, and the road to
the palace was covered with the sick and dying, the
nabob one day came out of the palace, seated on his
elephant. His servants made no attempt to clear the
road, or to prevent the poor sufferers from being trodden
under foot. But the elephant, of his own accord, took
thought for this. He lifted some of them out of the
path with his trunk, and among the others he stepped
with so much care and caution that none of them, ultimately, were injured.

Mr. Forbes, in his "Oriental Memoirs," says of the
elephant on which he was accustomed to ride,—

"Nothing could exceed the sagacity, docility, and
affection of this noble quadruped. If I stopped to
enjoy a prospect, he remained perfectly immovable
until my sketch was finished. If I wished for ripe
mangoes, growing out of the common reach, he would
select the most fruitful branch, and breaking it off with

15

his trunk, give it to his driver to be handed to me. Any part of it given to himself he accepted with a salaam, raising his trunk three times above his head, and after expressing his thanks by a murmuring noise. If a bough obstructed the howdah in which I sat, he twisted his trunk round it and broke it off; and often would he gather a leafy branch as a fan, to wave around us with his trunk. He generally paid a visit to the tent-door during our breakfast, where he always received some sugar-candy or fruit. No spaniel could be more innocent or playful, or fonder of those who took notice of him, than this noble creature."

A female elephant which, some years since, was kept at the Duke of Devonshire's villa, at Chiswick, showed both great docility and much cleverness. She had a house of large dimensions, and a paddock of considerable extent. She was thus well provided for, and she was not backward in showing herself sensible of the kind treatment she received. When called by her keeper, she would instantly come out of her house, and take up a broom, ready to do his bidding in sweeping the grass or the paths. Or, if bidden, she would follow him with a pail or a watering-pot. Her reward was generally a carrot and some water. If a bottle of soda-water was given to her, she would hold it against the ground with her foot, while with her trunk she gradually twisted out the cork, then reversing the bottle so as to empty the contents into her proboscis, delivering the empty bottle to her keeper, and then discharging from her proboscis the soda-water into her mouth. Her affection for her keeper was most evident. If he was absent for more than a few hours, she became restless, and shewed her anxiety by cries. At last the English climate told upon her, and she died, at the age of twenty-nine years, of something resembling that which in human beings is called pulmonary consumption.

Mr. Wood says,—

" One of my friends insulted, in some way, one of the elephants at the Zoological Gardens,—I believe, by offering an apple and then taking it away; or some other slight. The elephant shewed no resentment, and my friend went away. He strolled round the gardens, and presently came back again, and was going towards the elephants, when one of the keepers told him to keep out of the way of the large elephant, for it had got a heap of stones and brick-bats ready for him, and if he did not want to have them on his head, he had better keep out of the creature's way. He inquired a little further, and found that the elephant had really piled up six or seven large bricks or stones in a convenient place, and would probably have sent them upon his head if he had not avoided showing himself."

At Travancore, in the year 1811, a lady living with her husband was astonished to see an elephant, unattended, march into the court-yard of the fort, carrying in his trunk a box, evidently of no small weight. This he deposited in a suitable place, and went his way. He soon returned with another, and a third, and a fourth; continuing till he had formed a pile, arranged in excellent order. It turned out that the Rajah of Travancore had died in the night, and the English resident, for greater security, had removed the more valuable part of his possessions to the fort adjacent; and the elephant performed the part of a gigantic and careful porter,— acting, from his own consciousness, with the greatest prudence and foresight.

In another case an elephant was frequently left in charge of a baby, while the mother went about some necessary avocations; and no nurse could have been more careful or watchful over the little creature. The child, with the usual restlessness of infancy, would soon begin crawling about and getting into danger, and the elephant would, with care and tenderness, lift it into a better position, tending it with almost a mother's care, and giving it back, at last, unharmed and safe.

In a similar case an elephant took such a fancy to a little child as never to be quite comfortable except when

the baby was within sight. The nurse would often take
the child out of its cradle, and place it near to the
elephant. At length he became so accustomed to this
that he would hardly eat his food unless the child was
near. When the baby slept, he would drive away the
flies; and if it was fretful, he would rock the cradle till
it fell asleep.

Daniel Wilson, Bishop of Calcutta, says that "an
elephant belonging to an engineer-officer in his diocese
had a disease in his eyes, and had for three days been
completely blind. His owner asked Dr. Webb, a physi-
cian intimate with the bishop, if he could do anything for
the relief of the animal. Dr. Webb replied that he was
willing to try on one of the eyes the effect of nitrate of
silver, which was a remedy commonly used for similar
diseases of the human eye. The animal was accordingly
made to lie down, and when the nitrate of silver was
applied, uttered a terrific scream at the acute pain which
it occasioned. But the effect of the application was
wonderful, for the sight was in a great degree restored,
and the elephant could partially see. The doctor was
in consequence ready to operate similarly on the other
eye on the following day. The animal, when he was
brought out and heard the doctor's voice, lay down of
himself, placed his head quietly on one side, curled up
his trunk, drew in his breath, like a human being about

to endure a painful operation, heaved a sigh of relief when it was over, and then, by motions of his trunk and other gestures, gave evident signs of wishing to express his gratitude. Nothing could more plainly show memory, understanding, and reasoning."

The two following stories are of a similar character :—

During one of the wars in India, many Frenchmen had an opportunity of observing one of the elephants that had received a flesh wound from a cannon-ball. After having been twice or thrice conducted to the hospital, where he extended himself to be dressed, he afterwards used to go alone. The surgeon did whatever he thought necessary, applying even fire to the wound ; and though the pain made the animal often utter the most plaintive groans, he never expressed any other token than that of gratitude to this person, who by momentary torments endeavoured to relieve him, and in the end effected his cure.

In the last war in India, a young elephant received a violent wound in its head, the pain of which rendered it so frantic and ungovernable that it was found impossible to persuade the animal to have the part dressed. Whenever any one approached, it ran off with fury, and would suffer no person to come within several yards of it. The man who had the care of it at length hit upon a contrivance for securing it. By a few words and signs he gave the mother of the animal sufficient intelligence of what he wanted. The sensible creature immediately seized her young one with her trunk, and held it firmly down, though groaning with agony, while the surgeon completely dressed the wound; and she continued to perform this service every day till the wound was healed.

Two very young elephants were brought from Ceylon into Holland. For the convenience of the transit it was found necessary to separate them—which was not effected without difficulty. From Holland they were conveyed to Paris, where a spacious dwelling had been prepared for them. It was divided into two compartments, which communicated with each other by a door. The whole was enclosed by a strong wooden paling. On their arrival, they were taken to this habitation—the male arriving first. Immediately on entering his dwelling, he began a close scrutiny, examining every beam and trying the strength of every fastening. When he came to the door which connected the compartments, he soon discovered that it was fastened only by a single sliding bolt, which he at once lifted up, opened the door, and went into the other compartment, where he ate his breakfast. Presently the female arrived, who had only been separated from him, months before, with great trouble. The joy with which the two now met was hardly to be

described. Running to each other, they uttered loud
cries of delight, blowing the air from their trunks with
a noise that shook the building. Their ears moved
rapidly; their trunks were intertwined, and then thrown
round each other, and then applied to the ears, and
carried to the mouth. The male showed his ecstasy
by shedding abundance of tears. When their delight
had thus spent itself, they were allowed to have their
dwelling in common ; and their evident attachment
interested all beholders.

We have not written a "Natural History," or the
hundredth part of one. Our object has merely been to
direct the reader's attention to a few facts in that history,
which are interesting in themselves, and have an obvious
bearing on questions which are often now-a-days
brought into discussion.

Our subject has been, the existence of mind, of soul,
and spirit, in the animal creation. And while we have
been compiling the preceding pages, a fresh testimony
has been given by a pen which has often, in past days,
dealt with these subjects.

The Duke of Argyll, in a recent paper in a London
periodical, gives three instances of "Instinct" which had
come under his notice during the preceding Autumn.

"A pair of birds, known as the Dipper, or water-ousel, had built their nest at Inverary, in a hole in the wall of a small tunnel, constructed to carry a rivulet under the walks of a pleasure-ground. The season was one of great drought, and the rivulet, during the whole time of incubation, and of the growth of the young in the nest, was almost entirely dry. One of the nestlings, when almost fully fledged, was taken out of the nest, by the hand, for examination; an operation which so much alarmed the others that they darted out of the hole, and ran and fluttered down the tunnel towards its mouth. At that point a considerable pool of water had survived the drought, and lay in the path of the fugitives. One of them stumbled into the pool. Never before could he have seen water. When he touched it, there was a moment of pause, as if the creature were surprised. Then, instantly, there seemed to wake within it the sense of its inherited powers. Down it dived; and the action of its wings under the water was a beautiful exhibition of that peculiar adaptation which belongs to the wings of most of the diving-birds. The young bird was lost to sight among some weeds, and so long did it remain under water that I feared it must be drowned. But in due time it re-appeared, and was recaptured and replaced in the nest.

"Later in the season, I observed, on a secluded lake, a Dun-diver, or female of the red-breasted Mergauser, with her brood of young ones. On giving chase, we soon found that the young ones, though not above a fortnight old, had such extraordinary powers of swimming and diving that it was almost impossible to capture them. The distances they went under water, and the unexpected places in which they emerged, baffled all our efforts. And yet it was but two weeks before that time when these little things had been, each of them, coiled inside the shell of an egg !

"A third case was of a different kind. Walking along the side of a river, I came suddenly upon a wild-duck, whose young were just out. Springing from under the bank, she fluttered out into the stream, with loud cries, and with all the struggles, the flapping, and the wriggling of a badly-wounded bird. When she found that she could not induce us to follow her, she made resounding flaps on the surface of the water;—all these efforts having the same object—to draw our attention away from her young."

The Duke then argues, with great force, that there was nothing in either of these cases which could have been learnt, or gained by experience. Instinct, an intuitive knowledge or perception born with these creatures, was the sole cause of these movements or actions.

It is useful, and serves to excite a fresh interest in the question, thus to be reminded of old facts by new proofs and instances. But such incidents as these are familiar to all students of Natural History; and it is almost like advancing fresh evidence in favour of the Solar System, to give us proofs now-a-days of the reality and wonderful character of that "Instinct," which is inherent from the moment of birth, in a vast number of the creatures that we find, both on the land and in the water.

But can we be content with perceiving and recognising such a fact as this, without involuntarily going forward to certain natural conclusions? From our infancy we have been accustomed to the obvious question, "He that planted the ear, shall He not hear?—He that formed the eye, shall He not see?"

It has been admitted now for a long time past, by earnest thinkers, even of a sceptical turn of mind, that Paley's example of the watch affords a valid basis for an argument. A traveller, passing over a heath, finds a watch, in motion, which has been dropped, or left there by some preceding passenger. He takes it up, examines it, discovers its character, its purpose; and assuredly he will never be brought to believe that "it grew there," or came there by virtue of any "law" of "development," or of any other kind. Half an hour's reflection, or less, will bring him to an undoubting conclusion that this little machine, with its complex contrivances, its obvious purpose, and its regular performance, had a Maker.

That it *made itself*, or came into existence, slowly, by some law of development, he will see, in the course of a few minutes' reflection, to be mere talk without meaning. Even if it could be imagined, for a moment, that certain pieces of steel, of gold, and of other metals, could be strangely thrown together, it would be utterly inconceivable that, even in a thousand centuries, those pieces of metal could form themselves into a watch. And then, lifting up his eyes as the twilight and the darkness came on, and passing from the lowest to the highest—from the smallest to the largest—he would fix his thoughts on that other watch, that great Solar System, which proceeds on laws as intelligible and as well-known as the diminutive machine which he now holds within his fingers. The planets, with the sun as their centre, move much as the watch moves, but with an immenseness of area, and a certainty of action, to which the watch can only be compared as the eye of an ant may be compared with the eye of an archangel. Thus, then, whether the observer looks above or below, to the wondrously-constructed cell of the bee or web of the spider, or to the immensity of the higher heavens, with its countless zones of solar systems,—everything tells him alike of one wondrous, incomprehensible, and overwhelming Power, the Creator, Upholder, and Governor of all things, whose will called everything visible and invisible into being; and by whose will they continue, so long, and only so long, as His pleasure shall ordain.

Mind cannot proceed from anything less than mind. A mouse, or an oyster, renewed year by year, and proceeding, in endless succession, through ten thousand years, will never rise to be a parrot, or a pointer-dog. The simple story told to us in the first chapter of Genesis is not only to be received because it is the Word of God, but also as the only credible and the only intelligible account we have of the rise of " The Animal Creation." " *God made* the beast of the earth after his kind, and everything that creepeth upon the earth after his kind, and God saw that it was good."

The words of the Lord Jesus were not idle words; no one knew better than He, how absolutely true they were. " Consider the lilies of the field; I tell you that Solomon in all his glory was not arrayed like one of these." We try to appreciate and understand these words, and we soon discover that they are true, but that their truth surpasses our understanding. " There is not a fish that swims in the sea whose scales are not more beautifully woven than any texture ever fabricated by man. The variety, too, is endless. The fibres in the scale of a pike are quite different from those in the scale of a carp or a perch, while these latter equally differ from those of all other fishes. The dust on the wing of a butterfly proves, in the microscope, to be a number of exquisitely-formed feathers, exhibiting the most delicate and admirable arrangement in all their parts.   The moth, again, has a plumage wholly different from that of the butterfly ; and each species has feathers differently formed from all others.   The sting of a bee or of a gnat is found to be formed with surprising beauty, and with the most exact regularity.   There is a polish without flaw or blemish, ending in a point too fine to be discerned. The silkworm's web is found to be perfectly smooth and shining, and every thread equal and of astonishing fineness.   The spider's web· is still more delicate.   The minutest hair and fibre on a flea, a gnat, or a fly, is smooth and beautifully polished ; in fact, turn where we will, beauty, order, and perfection meet our gaze. Myriads of creatures fill the air, the sea, and the forest, whose bodies, or wings, or limbs, are embellished as with jewels : they are coloured with azure, green, red, or vermilion ; yet in all there is nothing gaudy or incongruous.   Truly must the student of nature exclaim with Eliphaz, ' God doeth great things and unsearchable; even marvellous things without number.' " (" Essays on the Bible," Essay VI.)

Still, mind claims pre-eminence over matter. Re· garded merely as a machine, the man, the eagle, the humming-bird, the moth—all are "fearfully and wonderfully made."   But that which dwells within the outward frame is still more wonderful, and much more interesting.

"God created man in His own image, and said, Let them have dominion over the fish of the sea, and over the fowl of the air, and over the cattle, and over every creeping thing that creepeth upon the earth." This dominion was not merely that of the strong over the weak. The event has proved that the "dominion" was not a supremacy of force, but of something higher.

Those creatures which are most useful and most interesting to man—such as the horse and the dog—show, in the most explicit manner, their willingness, their desire, to be his servants; and, as far as they may be, his friends. Others, such as the ox and the sheep, resign themselves willingly to the condition in which God has placed them, and readily occupy the pastures which man prepares for them. A third class—the birds—like the first, gladly occupy places in man's family, and rejoice to receive his notice, and to be valued by him.

We ought not to overlook these facts, or to disregard them. The preceding pages have been filled with a large number of instances of them; and they all tend one way. They all teach us to exclaim, with the Psalmist, "O Lord, how manifold are Thy works! in wisdom hast Thou made them all; the earth is full of Thy riches. . . Thou sendest forth Thy Spirit, they are created; and Thou renewest the face of the earth. The glory of the Lord shall endure for ever; the Lord shall rejoice in His works."

CHAPTERS ON ANIMALS.

By PHILIP G. HAMERTON.

In Post 8vo. with Twenty Etchings by Veyrasset and Bodmer.

Price 12s. 6d.

DOG-LIFE.

NARRATIVES EXHIBITING INTELLIGENCE, SYMPATHY, AND AFFECTION.

In Crown 8vo. with Engravings after Landseer.

Price 5s.